S0-EMA-269

Whitworth College Library
0 0068 00118352 1

UNITED STATES
DEPARTMENT OF COMMERCE
Jesse H. Jones, Secretary

BUREAU OF THE CENSUS
J. C. Capt, Director
Philip M. Hauser, Assistant Director

UNITED STATES
LIBRARY OF CONGRESS
Archibald MacLeish, Librarian

REFERENCE DEPARTMENT
David C. Mearns, Director

GENERAL CENSUSES AND VITAL STATISTICS IN THE AMERICAS

An annotated bibliography of the historical censuses and current vital statistics of the 21 American Republics, the American Sections of the British Commonwealth of Nations, the American Colonies of Denmark, France, and the Netherlands, and the American Territories and Possessions of the United States

Prepared under the supervision of
IRENE B. TAEUBER
Chief, Census Library Project

GOVERNMENT PRINTING OFFICE, WASHINGTON, 1943
Republished by Blaine Ethridge--Books, Detroit, 1974

ALSO PUBLISHED BY BLAINE ETHRIDGE

Bibliography of Selected Statistical Sources of the American Nations. A Guide to the Principal Statistical Materials of the 22 American Nations. Prepared by the Inter American Statistical Institute. Blaine Ethridge reprint of: Washington--Pan American Union, 1947, xvi + 689 pp. (Updated by quarterly supplements in *Estadistica.*)

Acclimatization in the Andes. By Carlos Monge. New preface for this edition by Paul T. Baker, International Coordinator, International Biological Programme High Altitude Studies. Blaine Ethridge reprint of Baltimore--The Johns Hopkins Press, 1948, xix + 130 pp.

WRITE FOR COMPLETE CATALOG

BLAINE ETHRIDGE--BOOKS
13977 Penrod Street, Detroit, Michigan 48223

This book was reproduced from a copy in the
Sociology and Economics Department
of the Detroit Public Library

Library of Congress Catalog Card Number 73-81474
International Standard Book Number 0-87917-036-0

PREFACE

The development of censuses and vital statistics in recent decades has resulted in a rapid increase in both the variety and the quantity of data published by the governments of the countries of the Americas. At the same time, the expansion of research and administration into the international field has made it essential that private scholars and governmental personnel have ready access to the demographic statistics of all countries. In 1938 a group of population students in the United States began the discussion of possible ways by which collections of the censuses and vital statistics of the world might be completed and made available for research and operational uses. Their deliberations resulted in a plan which was endorsed by the Eighth American Scientific Congress in the following resolution:

WHEREAS: Since the collection in a single depository library of all population, census, and statistical publications throughout the world would be of invaluable assistance to the students of population and national problems in the American Republics,
The Eighth American Scientific Congress
RESOLVED: To endorse through the Inter American Statistical Institute the creation of a special census unit in the Library of Congress of the United States and that it urge the governments of this Western Hemisphere to cooperate in this enterprise.

In accordance with this resolution, a Census Library Project was initiated in the Library of Congress in the fall of 1940 as a cooperative project of the Bureau of the Census and the Library of Congress. The advisory committee to the Library of Congress on the Project consists of the following persons, representing the agencies indicated: Luther H. Evans, Library of Congress; Philip M. Hauser, Bureau of the Census; Stuart A. Rice, Inter American Statistical Institute; Richard O. Lang, American Statistical Association; Frank W. Notestein, Office of Population Research, Princeton University; and Frank Lorimer, Population Association of America.

It became apparent during the early period of the operation of the Census Library Project that the completion of the collections was only one aspect of the problem of making international census and vital statistics materials available to students in the Americas. Consequently, the scope of the Project was expanded to include the following functions: (1) Cooperation in the completion of the collections of the Library of Congress in the fields of census and vital statistics for all countries, (2) the compilation of analytical bibliographies to facilitate the use of the collections, (3) the provision of reference and consultant services in the population field, and (4) the execution of such special studies as should be deemed advisable by the sponsoring agencies.

The exigencies of war necessitated an immediate shift of emphasis from this long-range program to the servicing of the immediate needs of governmental agencies. National censuses constitute the basic source of statistical information on peoples, economies, and cultures, as well as the primary analytical data in such diverse fields as manpower potential, racial heterogeneity, and industrial distribution. Changes in birth and death rates reflect the changing impact of war on the morale, health, and vitality of peoples, while the level of mortality and the specific causes of death reflect health hazards in various parts of the world.

The sudden expansion of the need for information and analysis of these and many other types of problems made questions of the adequacy, availability, and timeliness of census and vital statistics data one of immediate practical concern. A previous publication of the Project, "Recent censuses in European countries, a preliminary list, 1942," indicated that systematic bibliographical coverage was a primary first step in meeting the emergency situation.

It is hoped that the present bibliography relating to the population censuses and current vital statistics of the Americas will be not only an aid in the immediate problems of locating source materials but also a contribution to the development of statistical cooperation and intellectual understanding among the peoples of the Americas. It has been compiled by a limited staff, subject to the continuous pressure of immediate tasks. It is offered as an initial attempt at a comprehensive picture of the official sources of demographic statistics in the Western Hemisphere, not as a definitive study of the resources of any one area. Such definitive studies can be made only in some future period when the leisurely pursuits of scholarship are again possible, and even then, they must be made individually in each country, preferably by students from the country itself. In the meantime, it is hoped that this bibliography of the source materials for all the nations will not only facilitate the use of available statistical data but also stimulate the exchange of information and publications among the various nations. If these results are attained it will represent a step in the development of that demographic center in the Library of Congress which was envisioned in the resolution of the Eighth American Scientific Congress.

This study of the demographic publications of a hemisphere represents the cooperative activity of many persons and agencies. The compilation of check lists of the censuses of the countries of Latin America was started by Jesse H. Shera, the first supervisor of the Project. Responsibility for the arduous task of compilation of sources for the enlarged study was shared by Henry Dubester, Mary M. Kieron, Muriel McKenna, and Margaret Stone, past or present members of the Project staff, and Beatrice Sattler of the Office of Population Research of Princeton University.

Special acknowledgment should be made of the continued encouragement of Luther H. Evans, Chief Assistant Librarian of the Library of Congress; Philip M. Hauser, Assistant Director of the Bureau of the Census; Frank W. Notestein, Director of the Office of Population Research of Princeton University; and Halbert L. Dunn, Chief, Vital Statistics Division of the Bureau of the Census, who was a member of the Committee on the Census Library Project from the time of its inception until recently. The cooperation of many members of the staffs of both the Bureau of the Census and the Library of Congress is also acknowledged, although the staff of the Census Library Project alone should be held responsible for errors, whether of commission or of omission.

IRENE B. TAEUBER,
Chief, Census Library Project.

CONTENTS

PART I. THE AMERICAN REPUBLICS

PART II. AMERICAN SECTIONS OF THE BRITISH COMMONWEALTH OF NATIONS

PART III. AMERICAN COLONY OF DENMARK

PART IV. AMERICAN COLONIES OF FRANCE

PART V. AMERICAN COLONIES OF THE NETHERLANDS

PART VI. AMERICAN TERRITORIES AND POSSESSIONS OF THE UNITED STATES

INTRODUCTION

In 1579, forty years before the Pilgrims landed at Plymouth Rock, Philip II of Spain ordered a survey of the physical, social, and economic resources of all the territories subject to the Crown of Spain. Whether or not this survey constituted the first census in the Americas, or whether it was a census at all, is immaterial. It emphasizes the antiquity of that search for quantitative data which has been so omnipresent a characteristic of the colonies and nations which developed on the new continents. The extent to which the general realization of the need for accurate and detailed information on resources, economy, and people led to the development of national censuses and vital statistics differed widely from country to country, and even from period to period within the same country. The development of demographic statistics in each country has been related intimately to the general social, economic, and political history of the country, and influenced by the vicissitudes of particular events and the presence or absence of outstanding individuals especially interested in the field.

It is only within a recent period that any considerable proportion of the political areas of the Americas have had any permanent census staffs, or even continuing plans for the census taking. In addition, there has been little cooperation or coordination between countries in the time, methods, techniques, or scope of census enumeration and publication. The problem of locating source materials for the study of the population history and the population problems of the various regions of the Americas thus becomes quite difficult. There have been many attempts to grapple with this problem of discovering source materials and making them accessible to the scholars of the various nations. *The Economic Literature of Latin America*, compiled by the staff of Harvard University's Bureau for Economic Research in Latin America, includes a brief résumé of the census history of each of the nations of Latin America.[1] *The Handbook of Latin American Studies*, initiated in 1935 by the Committee on Latin American Studies of the American Council of Learned Societies, includes annotated citations to the demographic publications of the various countries.[2] Notes on the demography of the countries of Latin America are published regularly in the *Boletín de la Oficina sanitaria panamericana*. In addition, this Bureau issues compilations of the vital statistics of the various countries.[3] *Population Index*, the quarterly journal issued by the Office of Population Research of Princeton University and the Population

[1] Harvard University. Bureau for Economic Research in Latin America. *The economic literature of Latin America, a tentative bibliography*. Compiled by the staff of the Bureau of Economic Research in Latin America, Harvard University, Cambridge, Harvard University Press, 1935–1936. Vol. I. Latin America, Argentina, Bolivia, Brazil, Chile, Colombia, Ecuador, Paraguay, Peru, Uruguay, and Venezuela. Vol. II. Latin America. Mexico and Central America, Mexico, Guatemala, Honduras, Salvador, Nicaragua, Costa Rica, Panama, the West Indies, Cuba, Puerto Rico, Haiti, Dominican Republic, Panama Canal.
Z 7164.E2 H36

[2] The following citation refers specifically to the last issue covered for this bibliography: Burgin, Miron, ed. *Handbook of Latin American studies: 1940*. No. 6. A selective guide to the material published in 1940 on anthropology, archives, art, economics: . . . Edited for the Committee on Latin American Studies of the American Council of Learned Societies. Cambridge, Harvard University Press, 1941. 570 pp.
Z 1605.H23 1940

[3] Oficina sanitaria panamericana. "Demografía de las Repúblicas Americanas." Reimpreso del *Boletín de la Oficina sanitaria panamericana*, enero, 1940. Oficina sanitaria panamericana, Publicación No. 142 Washington, 1940, pp. 15–30.

Association of America, includes citations to the current official and unofficial population literature of the Americas.[4] The report on the *Statistical Activities of the American Nations, 1940,* edited under the auspices of the Temporary Organizing Committee of the Inter American Statistical Institute, includes for each of the cooperating countries a report on statistical activities written by one or more statisticians of the country itself.[5] In addition, the editors presented summary statements for each country on the statistical organization, the national population censuses, the principal government agencies which compile statistics, and the principal official serial statistical publications (latest issue). Finally, the *InterAmerican Statistical Yearbook, 1942,* prepared under the direction of Raul C. Migone and sponsored by the Argentine Commission of High International Studies, contributed directly to the present bibliography in that it facilitated the process of checking on the completeness of coverage and on the sources available.[6]

The compilation of a systematic bibliography of the official population literature of two continents within a limited time period would have been impossible without heavy reliance on these prior publications. Particularly valuable were the reports on the statistical systems and statistical activities of the various countries published in the *Statistical Activities of the American Nations.* The editorial staff was also peculiarly fortunate in having free access to the master author files of *Population Index,* which are housed in the offices of the Census Library Project.

This report on the *Population Censuses and Current Vital Statistics of the Americas* consists of a separate report for each of the 21 American Republics, Canada, Newfoundland and Labrador, and the colonies and territories of the British Commonwealth of Nations, Denmark, France, the Netherlands, and the United States. Insofar as the nature of the basic data permitted, each report consists of 3 sections: a historical note; a list of all the published national population censuses located, with annotation of the most recent one; and an annotated list of the sources for current vital statistics and population estimates. In many cases, a fourth section was added on other current national population statistics, including especially life tables and statistical atlases and compendia.

Obviously, it was impossible to outline any general *a priori* plan which could be followed to permit uniform and comparable coverage in a bibliography of the publications in countries as diverse as Argentina, Paraguay, French Guiana, Canada, and Greenland. To attempt to achieve complete comparability in the reports for all countries would have been to sacrifice the guiding principle of service and utility for the unattainable goal of uniformity. Actually, the coverage of past censuses is least complete for countries with a long history of census taking and a massive published literature. The official statistical organizations of such countries usually provide check lists of their own publications. It is probable that either they or other individuals or organizations within the country have compiled and published detailed histories and bibliographies of the statistical development and publications of the country.

The policy with reference to the inclusion of provincial census and vital statistics has also differed from country to country. Provincial or city censuses and vital statistics are of unquestioned importance in any country; they become of para-

[4] Office of Population Research of Princeton University and Population Association of America, Inc. *Population Index.* Vol. I, 1935 to date. Last issue: Vol. IX, No. 2. Princeton, New Jersey, 1943.
Z 7164.D3 P83

[5] Inter American Statistical Institute. Temporary Organizing Committee. *Statistical activities of the American nations, 1940. . . .* Edited by Elizabeth Phelps. Washington, Inter American Statistical Institute, 1941. xxxi, 842 pp.
HA 175.S75

[6] Migone, Raul C., Director, with Aberastury, Marcelo F., Assistant Director, and Fuente, Emilio, and Iturraspe, Jorge E., coauthors. *Inter American statistical yearbook.* Issued under the auspices of the Argentine Commission of High International Studies. New York, The MacMillan Co.; Buenos Aires, El Ateneo; Rio de Janeiro, Freitas Bastos y cía, 1943. 1066 pp.

mount importance only in countries in which there are no current and detailed national censuses or vital statistics. Similarly, there has been no rigid rule on the treatment of censuses other than population censuses when taken separately and not as an integral part of a general national census which included population. In most cases, inclusion has been limited to the most recent census, with no attempt at comprehensive historical guides in such diverse fields as industry, agriculture, manufacturing, distribution, employment, and unemployment.

The problem of what constitutes a census has been an ever-recurring one to which no satisfactory answer has been found. Scientific studies of the validity of population censuses or vital statistics in any particular country necessitate not only formal analysis of the published results for the country as a whole and its various regions, but also knowledge of the process of enumeration and tabulation used in each case, and the particular physical, economic, or political factors influencing the completeness and accuracy of the schedules secured for the various regions and classes. It is probable that such a comprehensive study of validity is the task of the census and vital statistics staffs of the individual countries. Certainly it is not a task to be undertaken with a limited staff for all the countries and colonies of two continents. In any event, the purpose of the present study has not been that of the evaluation of the inclusiveness and quantitative validity of the various censuses, but rather that of offering to research students a guide to the quantitative or partially quantitative data available in the official publications of the various countries. Hence the censuses included in the present compilation are essentially those enumerations or partial enumerations which are accepted as censuses by the statistical bureaus of the various countries. Evaluative comments, where they are included, are based on the comments and analyses included in the published reports of the country itself.

The form and contents of this bibliography reflect one of the primary motives in its compilation—service to the war agencies of the Government of the United States. A careful coverage of the materials in the Library of Congress was followed by searches for missing publications in the Columbus Memorial Library of the Pan American Union, the Army Medical Library, and the libraries of the Pan American Sanitary Bureau, the Department of Commerce, the Bureau of the Census, and the Inter American Statistical Institute. Call numbers are included for Library of Congress holdings, or, for the materials not yet classified, the location within the Library. If publications were located elsewhere in Washington, the name of the library is given.[7]

[7] The following abbreviations are used: *Govt. Publ. R. R.*—Government Publications Reading Room of the Library of Congress; *Pan. Am. Union*—Columbus Memorial Library; *Pan. Am. San. Bur.*—Library of the Pan American Sanitary Bureau; *Dept. of Comm.*—Department of Commerce Library; *Bur. of Cen.*—Bureau of the Census; *Inter Am. Stat. Inst.*—Inter American Statistical Institute.

Part I
THE AMERICAN REPUBLICS

ARGENTINA

Historical

A long history of legislation and agitation preceded the actual execution of the first national census of Argentina in 1869.[1] A census of all the provinces of the viceroyalty was ordered in 1810, but only that for Buenos Aires was taken, and even these results were not published. In 1813 the constitutional assembly of the Council of the Revolution issued a decree providing for a census, which was to include information on classes, marital status, place of origin, age, and sex. This census was not taken.

The Provisional Regulations of 1815 and 1817, and the Constitutions of 1819 and 1826 stressed the need for a general census of population in order to determine the number and distribution of deputies. The Unitarian Constitution of 1826, Article 12, authorized a census every 8 years, but it was ruled subsequently that this meant a census not oftener than every 8 years. No censuses were taken during this period. The present constitution, that of 1853, provides that the number of deputies be apportioned according to provincial populations.

There have been only three national censuses in Argentina history, those of 1869, 1895, and 1914.[2] The first census consisted of a simple count of the population. The second included some economic data, while the third and last included economic and agricultural data as well as detailed population tabulations. Entirely aside from any questions of its scope and validity, however, it portrays the Argentina of 1914, not that of 1943.

In 1939 the Chamber of Deputies of the National Congress voted to authorize a census of population in 1940. However, parliamentary proceedings were not completed, so the action of the lower house became only a public recognition of the need for a census.

Apart from the three published censuses of Argentina, annual estimates of the population as of December 31 have been prepared by the *Dirección general de estadística*. The estimates, which are the result of adjusting births, deaths, emigration, and immigration to the latest census count, were published first in 1926, and included the period from 1910 to 1925.[3] These and subsequent estimates appear in the reports of the *Dirección general de estadística*, Series D, supplemented since 1931 by annual mimeographed releases.[4] The method employed in obtaining the estimates is explained in the 1910–25 report.

[1] Comisión nacional del censo. *Tercer censo nacional levantado el 19 de junio de 1914. . . .* Buenos Aires Talleres gráficos de L. J. Rosso y cía, 1916. Tomo 1, pp. v–vi. Discussion by Alberto B. Martínez, president of the census commission.

[2] Dieulefait, Carlos E. *Estadística censal y estadística administrativa argentinas.* Rosario, Talleres gráficos Pomponio, 1935. 94 pp. (Universidad nacional del Litoral. Facultad de ciencias económicas, comerciales y políticas. Instituto de estadística.) HA 37.A7 5R6

See also Ibid., "Las actividades estadísticas de la Argentina." Pp. 3–25 in: Inter American Statistical Institute. *Statistical Activities of the American Nations.* Washington, 1941.

[3] Dirección general de estadística. *La población y el movimiento demográfico de la República Argentina en el período 1910–1925.* Informe No. 20, Serie D, Demografía, No. 1. Buenos Aires, G. Kraft Ltda., 1926. HA 943.A36

Other reports in this series containing population estimates are: No. 3, 1937–1936, published 1938. No. 4, 1938–1937, published 1939. No. 6, 1939–1938, published 1940. No. 7, 1941–1940, published 1942. HA 945.A5

[4] Dirección general de estadística. *La población de la República Argentina el 31 de diciembre de 1931.* Buenos Aires, 1932, to date. HA 943.A34

Information concerning the population of Argentina available from the national censuses and official estimates may be supplemented by provincial and city censuses. Typical of such enumerations are the census of the Province of Mendoza, 1942, and the four censuses of the city of Buenos Aires taken in 1887, 1904, 1909, and 1936. There has been considerable agitation for the coordination of the work of the provincial and municipal bureaus, with plans for an annual résumé by the *Dirección general de estadística* which would systematize national, provincial, and urban statistics. No such centralization has yet been achieved.

NATIONAL POPULATION CENSUSES

CENSUS OF 1869

Superintendente del censo.

Primer censo de la República Argentina. Verificado en los días 15, 16 y 17 de setiembre de 1869, bajo la dirección de Diego G. de la Fuente, superintendente del censo. Buenos Aires, Impr. del Porvenir, 1872. lx, 746 pp. HA 942 1869

CENSUS OF 1895

Comisión directiva del censo.

Segundo censo de la República Argentina, mayo 10 de 1895. Primeros resultados, mayo 10 de 1896. . . . Buenos Aires, J. Peuser, 1896. 80 pp. HA 942 1895

Second recensement de la République Argentine, 10 mai 1895. Résumés définitifs. Population nationale et étrangère, urbaine et rurale. . . . Buenos Aires, Impr. de J. A. Alsina, 1897. 39 pp. HA 942 1895b

Segundo censo de la República Argentina, mayo 10 de 1895. Buenos Aires, Taller tip. de la Penitenciaría nacional, 1898. 3 vol. HA 942 1895a

Tomo I. Capítulos I–II. Territorio. xxiv, 662 pp.
Tomo II. Capítulo III. Población. cxciii, 709 pp.
Tomo III. Capítulos IV–XV. Censos complementarios. cclv, 492 pp.

CENSUS OF 1914

Comisión nacional del censo.

Tercer censo nacional levantado el 1° de junio de 1914. Ordenado por la ley n° 9108. Buenos Aires, Talleres gráficos de L. J. Rosso y cía, 1916–1919. 10 vols. HA 942 1914

Tomo I. Antecedentes y comentarios.
Tomo II. Población. [Population for provinces and territories, by citizenship status. Increase, 1895–1914. Population by urban-rural residence, sex, and nationality, for provinces and territories. Population by sex and nationality, 1895–1914. Ethnic composition, 1895–1914. Urban-rural composition, 1895–1914. Naturalization: Civil status, nationality, sex, and age, provinces.]
Tomo III. Población. [Age, nationality, and sex, provinces and territories. Age by nationality. General résumé, age. Median age. Population and culture. School population. Population 6–14 years of age, 1895, 1909, and 1914.]
Tomo IV. Población. [Fertility. Married women by age and number of births. Infirmities. Occupations. Urban population.]
Tomo V. Exploraciones agropecuarias.
Tomo VI. Censo ganadero.
Tomo VII. Censo de las industrias.
Tomo VIII. Censo del comercio. Fortuna nacional. Diversas estadísticas.
Tomo IX. Instrucción pública. Bienes del estado.
Tomo X. Valores mobiliarios y estadísticas diversas.

OTHER NATIONAL CENSUSES

Comisión nacional del censo agropecuario.

Censo nacional agropecuario, año 1937. Buenos Aires, G. Kraft Ltda., 1940. 2 vol. HD 1861.A6 1937a

Pte 1. Economía rural.
Pte 2. Economía rural. Industrias derivadas. Varios.

Two preliminary reports were issued on this census in 1938: 1. *Las plantaciones de caña de azúcar.* Resultados generales. Cifras provisionales. 41 pp. 2. *Resultados generales.* Cifras provisionales. 23 pp. *See also: Censo nacional agropecuario.* Compendio, 1937. Buenos Aires, 1940. 110 pp.

Ministerio de hacienda. Comisión nacional del censo industrial.
Censo industrial de 1935. Buenos Aires. Talleres de la S. a. Casa Jacobo Peuser, Ltda., 1938. xliii, 750 pp.
Introducción. Parte I. Estadísticas generales . . .: Parte II. Estadísticas por capital Federal, provincias y territorios. Parte III. Estadísticas por rubros de industria. HC 171.A64 1935
Reports of the Industrial Census of 1940, published in 1942, are now available.

RECENT PROVINCIAL CENSUSES

[Official sources indicate that the following censuses have been taken: Federal Capital, Oct. 22, 1936. Province of Buenos Aires, Dec. 18, 1938. Province of Mendoza, June 24, 1942. Government of the Chaco, Jan. 23, 1934. Government of La Pampa, Oct. 3, 1935. *See:* Dirección general de estadística. *La población y el movimiento demográfico de la República Argentina en los años 1941 y 1940.* Informe No. 89, Serie D, No. 8. Demografía. Buenos Aires, 1942. 64 pp.]

MENDOZA

Mendoza. Gobernador.
Mensaje al inaugurarse el período ordinario de sesiones de la H. Legislatura. Mendoza, junio de 1942. J 202 M4 N15
The plans for the execution and tabulation of the 1942 census are described, pp. 265–272.

NATIONAL TERRITORIES

Argentina. Dirección general de territorios nacionales.
Censo de población de los territorios nacionales, República Argentina, 1912. Buenos Aires, Imp. G. Kraft, 1914. 370 pp. HA 943.A5 1912

Argentina. Ministerio del interior.
Censo general de los territorios nacionales, República Argentina, 1920. Tomo I. La Pampa, Misiones, Los Andes, Formosa y Chaco. Buenos Aires, Establecimiento gráfico A. de Martino, 1923. 480 pp. HA 943.A4
For each territory this volume includes: *Censos de población, ganadería, agricultura, industrias, comercio, vehículos.* The population census includes distribution, nationality, sex and age, education, population of school age, civil status, vaccination, and occupation.
Volume II, which presumably has the data for the remaining territories, was not located.

SANTA FÉ

Santa Fé. Ministerio de instrucción pública y fomento. Comisión central del censo escolar.
Cuarto censo general de la población en edad escolar de la Provincia de Santa Fé realizado el 14–X–1937. Santa Fé, 1941. 150 pp. Private Library.
This census includes boys 6 to 14 years of age and girls 6 to 12. Similar censuses were taken in 1912, 1918, and 1925, although there has been no general population census of Santa Fé since 1914. Detailed tables are presented showing by departments and zones the number of children of school age attending and not attending school, by sex and type of residence. Detailed information on the organization and execution of the census is included.

CITY POPULATION CENSUSES

BUENOS AIRES

Buenos Aires. Comisión técnica encargada de realizar de cuarto censo general.
Cuarto censo general, 1936. Población. 22–X–1936. Buenos Aires, Talleres gráficos de Guillermo Kraft, 1938–1940. 4 vol. HA 959.B94 1936
Tomo I. Informe preliminar. 1938. 430 pp.
Tomo II. Masculinidad. Lugar de nacimiento. Alfabetismo. 1939. 463 pp.
Tomo III. Estado civil. País de matrimonio. Religión. 1939. 413 pp.
Tomo IV. Fecundidad, familias. 1940. 393 pp.
Vol. I contains a general introduction to the population census, and a description of the plans, techniques, forms and procedures of this census. It also presents tables and graphs on de facto and de jure populations by age, sex, and place of origin, with comparative data from the census of 1914. Vol. II includes tables and graphs on age composition, origin, sex ratios, place of birth by age and sex, and literacy by age. Vol. III includes de jure population by age, marital status

and origin; marital status by place of birth; religion by place of origin, age, sex, and literacy. Vol. IV contains tables and graphs on fertility and the family.

ROSARIO

Rosario. Consejos escolares electivos. Comité de coordinación.
Censo infantil, 1934, 0–14 años, levantado el día 24 de agosto. Rosario, 1935. 203 pp. L 293 R6A5 1934

SANTA FÉ

Santa Fé. Dirección de estadística municipal.
Censo municipal de la población de Santa Fé levantado el 29 de julio de 1923. Santa Fé, Talleres gráficos "La Unión" de Ramón Morales, 1924. 284 pp. HA 959.S35 1923
Density, rural-urban distribution, sex, age, nationality, marital status, education, religion, physical defects, vaccination, orphans, fertility, and occupation.

CURRENT NATIONAL VITAL STATISTICS

(Including Population Estimates)

Centro de estudiantes y colegio de graduados.
"Reseña de la vida económica Argentina durante el año 1940." *Revista de ciencias económicas 29 (Serie II, 239); 210–226.* Suplemento extraordinario, June 1941. HB9.R37
The section devoted to population and labor covers vital statistics, population increase, and migration, 1935–1940. Issued annually.

Departamento nacional de higiene. Sección demografía y geografía médica.
Anuario demográfico del año 1936. Natalidad, nupcialidad y mortalidad. Año X. Buenos Aires, 1940. 143 pp. HA 943.A32
Births are tabulated by months, legitimacy status, and nationality; marriages by months, age, nationality and marital status; deaths by months, occupation, nationality; and cause of death by months, region, and age. The data for each province and territory include deaths by cause.
Publication irregular.

Dirección general de estadística.
Clasificación estadística de las causas de las defunciones. Nomenclatura internacional de 1938. Informe No. 77, Serie D, No. 5, Demografía. Buenos Aires, 1940. 38 pp. HA 943. A36

El comercio exterior Argentino en 1941 y 1940 y estadísticas económicas retrospectivas. Boletín No. 229. Buenos Aires, 1942. 273 pp. HF 159.A5
The section, "Demografía," summarizes data from the national censuses of 1869, 1895, and 1914, and the territorial censuses of 1895, 1905, 1912, 1914, and 1920. Annual population estimates are presented, 1914 through 1941. National vital statistics are summarized for the period from 1910 through 1941, provincial vital statistics for the period from 1931 through 1941. Migration statistics are summarized, 1857–1941.
Even-numbered volumes, covering the first half of a year, contain only commercial data. Odd-numbered volumes are yearly summaries and include a demographic section.

La población de la República Argentina el 31 de diciembre de 1941. Cifras provisionales. Buenos Aires, 1942. 18 pp. HA 943.A34
This annual release presents estimated population of provinces, migration, and vital statistics.

La población y el movimiento demográfico de la República Argentina en los años 1941 y 1940. Informe No. 89, Serie D., No. 8. Demografía. Buenos Aires, 1942. 64 pp. HA 943.A36
Chapter I gives estimates of the population by provinces up to Dec. 31, 1941. Chapter II presents vital statistics for 1941. Chapter III gives detailed tables on births, deaths, marriages, and infant mortality, with breaks for age, sex, legitimacy status.
Issued annually.

CURRENT PROVINCIAL VITAL STATISTICS

(Including Population Estimates)

BUENOS AIRES

Buenos Aires. Dirección general de estadística.

Anuario estadístico. Año 1936. La Plata, Taller de impresiones oficiales, 1937. 224 pp. HA 958.B915

The section, "Demografía," pp. 11–48, contains vital statistics and population estimates by minor divisions as of Jan. 1, 1937.

The 1938 issue appears to be the last one available in Washington. There is no information on further publication.

Boletín de la Dirección general de estadística de la Provincia de Buenos Aires 33 (255): 1091–1179. Jan., 1932. La Plata, Talleres de impresiones oficiales, 1932. HA 958.B85

This issue presents vital statistics for the last part of 1931.

CÓRDOBA

Córdoba. Dirección general de estadística.

Anuario estadístico. Año 1938. Córdoba, Talleres gráficos de la penitenciaría, 1942. 444 pp. HA 958 C8

Estimated populations and detailed vital statistics for 1938 are presented, pp. 236–362.

ENTRE RÍOS

Entre Ríos. Dirección general de estadística.

Síntesis estadística, año 1932. Parana, Imprenta de la provincia, 1933. HA 958.E6A3

The population section, pp. 4–17, gives estimated population and vital statistics. No later issues were located.

LA RIOJA

La Rioja.

For a summary statement on population, *see:* Sanches, Melchor B., *La Provincia da La Rioja; estudio físico, político y económico.* Buenos Aires, Talleres gráficos Argentinos L. J. Rosso, 1928. 246 pp. F 2956.S19

MENDOZA

Mendoza. Dirección general de estadística.

Anuario de la Dirección general de estadística de la Provincia de Mendoza . . . correspondiente al año 1936. Mendoza, Imprenta oficial, 1937. 282 pp. HA 958.M4A3

The section, "Extensión y población," gives population estimates and some data on aliens. The section on immigration summarizes trends, 1890–1936, and gives data on distribution, nationality and occupation for 1936. The section, "Demografía," summarizes vital rates for 1898–1936, analyzes trends in more detail for 1927–1936, and presents detailed data for 1936.

The 1936 *Anuario,* the last located, is in the Library of the Pan American Union.

Mendoza. Instituto técnico de investigaciones y orientación económica de la producción.

Estimación y análisis de la población de la Argentina con particular referencia a la de Mendoza. Economía y finanzas, Mendoza, Informe No. 2. Feb., 1940. Govt. Publ. R. R.

Includes data for the Chaco, La Pampa, Misiones and Formosa.

Tendencias demográficas de Mendoza. Natalidad, mortalidad, nupcialidad y población extranjera de 1914 a 1940. Informe . . . No. 3. Mendoza, 1940. 31 pp. Govt. Publ. R. R.

NATIONAL TERRITORIES

Argentina. Dirección general de estadística.

El movimiento demográfico en los territorios nacionales de la República Argentina en los años 1933 a 1941. Informe No. 91, Serie D, No. 9. Demografía. Buenos Aires, 1942. 162 pp. HA 943.A36

537846—43——2

The National Territories include the Chaco, Chobut, Formosa, La Pampa, Los Andes, Misiones, Neuquén, Río Negro, Santa Cruz, and Tierra del Fuego. Ch. I gives a general discussion of the method of obtaining the statistics. Ch. II presents summary data, 1933–1941. Ch. III is an analysis of the data for 1941. Ch. IV presents detailed vital statistics for each territory, 1941. Ch. V gives summary tables for territories, by months, 1941. Ch. VI contains summary tables by Civil Registers in each territory, 1941. Ch. VII summarizes data for the period 1933–1941 for each territory.

SALTA

Salta. Dirección general de estadística.

Boletín de estadística de la Provincia de Salta, año 1926. Salta, Imprenta oficial, 1927. HA 958.S2A3

Population estimates and vital statistics are included. No later issues were located.

SAN LUIS

San Luis.

For a summary statement on population, *see:* Gez, Juan E., *Geografía de la Provincia de San Luis.* Buenos Aires, Jacobo Peuser, Ltda., 1938–1939. 3 vol. F 2966.G364

"La población," Vol. II, Ch. IV, pp. 387–557, presents population and vital statistics data, primarily for the period 1931–1936. There are tables from the *Censo escolar nacional de 1931* which had not been published previously.

SANTA FÉ

Santa Fé. Dirección general de estadística.

Estadística demográfica, meses de enero a julio de 1934, comparados con datos de los años 1930–1933. Año 4, No. 8. Santa Fé, Casa de gobierno. 1935. 49 pp.

No later issues were located. HA 958.S8A45

Nacimientos, matrimonios y defunciones registrados durante los diez primeros meses del año 1930. Informe elevado al Ministerio de instrucción pública y fomento con fecha enero 20 de 1931. Año 1, No. 2. Santa Fé, Imp. de la provincia, 1932. 21 pp. HA 958.S84

SANTIAGO DEL ESTERO

Santiago del Estero. Dirección general de estadística, registro civil y trabajo.

Compendio estadística numérica de la Provincia de Santiago del Estero, 19–. HA 958.S86 A32

TUCUMAN

Tucuman. Oficina de estadística.

Anuario de estadística de la Provincia de Tucuman, correspondiente al año 1931. Tucuman, Impr. M. Violetto y cía., 1933. 494 pp. HA 958.T9A2 1931

The section, "Demografía," pp. 17–65, includes data on vital statistics, immigration and population estimates.

The 1934 issue is in the library of the Pan American Union. No later issue was located.

CURRENT CITY VITAL STATISTICS

(Including Population Estimates)

BUENOS AIRES

Buenos Aires. Dirección general de estadística municipal.

"Cálculo de la población funcional de la ciudad de Buenos Aires en el período 1536–1939." *Revista estadística municipal,* ciudad de Buenos Aires 52(622–624): 211–214. July–Sept., 1939. Govt. Publ. R. R.

Demografía. *Revista de estadística municipal de la ciudad de Buenos Aires,* 54(649–651): 327–371. Oct.–Dec., 1941 Govt. Publ. R. R.

A résumé of vital statistics for 1941, total and for minor areas, is followed by detailed tables on births, marriages, deaths, deaths by cause, and migration.

LA PLATA

La Plata.

Boletín de la Municipalidad de La Plata 33(290); 69. Jan.–June, 1941. Govt. Publ. R. R.

Summary vital statistics are presented semiannually.

ROSARIO

Rosario. Dirección general de estadística.

Anuario estadístico, 1940. Rosario, Emilio Fenner, S. R. L., 1942. 148 pp. HA 959.R7

Part IV. Population of Rosario, 1859–1940. Part V, Vital statistics.

Boletín estadística de la ciudad de Rosario. Vol. 12, 1938. HA 958.R714

The first section of this quarterly publication gives detailed vital statistics for the city. The issue for Sept., 1941, is in the Pan American Union Library.

SANTA FÉ

Santa Fé. Dirección de estadística municipal.

Anuario estadístico de la ciudad de Santa Fé. Año 1938, volume xxx. Santa Fé, Talleres gráficos Castellri Hermanos, 1940. xlvii, 205 pp. HA 959.S28

Section II, "Demografía," contains detailed vital statistics for 1938, with summary data for 1914–1938.

"Demografía." *Boletín de estadística municipal de la ciudad de Santa Fé 41(162); 2–12.* Jan.–March, 1942. HA 959.S3

Vital statistics are presented quarterly.

OTHER CURRENT NATIONAL POPULATION STATISTICS

Comas, Jorge M., and Goldenberg, Pedro.

Tabla de mortalidad de la República Argentina construida con los datos del 3er. censo nacional. Monografías de los alumnos del Instituto de biometría, No. 3. Faculdad de ciencias económicas de la Universidad nacional de Buenos Aires, 1936. 14 pp., 7 tables, 9 graphs. HA 13. B8 No. 3

Comité nacional de geografía.

Anuario geográfico argentino. Buenos Aires, 1941. 651 pp. Govt. Publ. R. R.

Ch. V, "Población," pp. 156–195, summarizes early estimates of population and the data of the censuses of 1869, 1895, and 1914, together with official estimates for Dec., 1939. (These provincial estimates are made by adding natural increase to the 1914 census population, assuming that migratory changes occur according to the observed coefficient of attraction between the censuses of 1895 and 1914.) The increase of total population is traced from 1797 to 1939. Summary data are also presented on age, 1869, 1895, 1914, and 1939 (estimated by Alezandro E. Bunge); religion; language; urban population, including Buenos Aires, 1536–1939; vital statistics, Federal Capital and provinces, 1910–1940; vital statistics of principal cities; historical and recent international migration by periods; demographic, economic, and nationality characteristics of migrants; and colonization.

BOLIVIA

Historical

Bolivia has had a long history of population censuses, although none appear to have been published in full except the last, that of 1900. A general census of the Provinces of Upper Peru was taken in 1796.[1] Population counts based on censuses taken in 1831, 1835, and 1846 are given in the first official publication of Bolivian population statistics, José María Dalencé's *Bosquejo estadístico de Bolivia.*[2] Population figures from censuses taken in 1831, 1835, 1845, 1854, and 1882 are reproduced in the first volume of the report of the Census of 1900.[3] According to the recent publications of the Bolivian Statistical Department, none of these censuses were complete or accurate enough to constitute a valid basis for population estimates.

The only census published in full for the entire country is that taken under the direction of the *Oficina nacional de inmigración, estadística y propaganda geográfica* in 1900. Lack of census experience, combined with difficulties of transportation and communication, resulted in incomplete returns for the enumerated areas. In addition, no attempt was made to cover the regions populated by uncivilized tribes. This population was estimated at 91,000 and added to the census population of 1,675,451.[4] Official population estimates from 1900 to the present have been made by taking this admittedly defective census as the base, and assuming a fairly fixed geometric rate of increase. From 1900 to 1933, the rate of increase for each province, and hence for the country as a whole, was assumed to be 1.75 percent per year. From 1933 to 1935, this rate was reduced to 1.0 percent per year because of the population losses of the Chaco War.

The *Dirección general de estadística* began to plan for a new census in 1938, but various factors operated to prevent its realization.

There have been no provincial censuses which might partially compensate for the lack of a recent national census. However, there have been at least two censuses of La Paz, one in 1909[5] and another in 1942.[6] Preliminary reports of the 1942 census indicate a population of 301,450, including 12,531 aliens and a floating population estimated at 14,353.[6]

---

[1] Pando Gutierrez, Jorge. "Las actividades estadísticas de Bolivia." pp. 33-57 in: Inter American Statistical Institute. *Op. cit.* HA 175.S75

[2] Dalencé, José M. *Bosquejo estadístico de Bolivia.* Chuguianca, Imp. de Sucre, 1851. 391 pp. (In New York Public Library.)

[3] Oficina nacional de inmigración, estadística y propaganda geográfica. *Censo general de la población de la República de Bolivia según el empadronamiento de 1° de septiembre de 1900.* La Paz, J. M. Gamarra, 1902. Tomo I, 29, xliv, 400 pp. *See also:* "Economic Literature of Latin America." Reproduced in: Inter American Statistical Institute. *Op. cit.*, p. 751. North reports six censuses between 1831 and 1900, and states that only the last was published. *See:* North, S. N. D. "Uniformity and cooperation in the census methods of the Republics of the American continent." pp. 53-79 in: *Trabajós del cuarto congreso científico, 1° Pan-Americano,* Vol. X. AS 4. P. 2. 1908

[4] Pando Gutierrez, Jorge, *op. cit.*, pp. 45-46.

[5] Comisión central del censo. *Censo municipal de la ciudad de La Paz, 15 de junio de 1909. . . .* La Paz, Tall. tip. lit. de José Miguel Gamarra, 1910. xiii, 80, ix pp. This report of the 1909 census includes figures from city or national censuses taken in 1675, 1796, 1831, 1845, 1886, and 1902, but no detailed information is given as to the nature of these early so-called municipal censuses.

[6] Unpublished data furnished by the Dirección general de estadística, Sección demográfica.

Vital statistics have been fully as inadequate as census materials. This inadequacy is admitted freely by the officials of the Bolivian Statistical Bureau, who attribute it primarily to the absence of a system of civil registration.[7] The last *Demografía* available, that for 1940, expresses the hope that the newly established Civil Register will permit the vital statistics for 1941 to be approximately 90 percent complete.[8]

Historically, statistics on the number of births have been based on the records of the churches. The statistics on the number of deaths have been based on the records of burial permits issued. The result is that vital statistics were not sufficiently complete to furnish a basis for estimating populations. Neither did they furnish a measure of the level of either fertility or mortality. In the introduction to the *Demografía*, 1939, it is pointed out that the birth rate for Bolivia, computed on the basis of the 1939 estimates, is 15.9, as compared with 34.5 for Chile, 23 for Argentina, 36 for Ecuador, 38.6 for Mexico, and 19.7 for Uruguay.[9] Internal analysis of the vital statistics corroborates the *a priori* assumption that the reported rate of 15.9 does not represent the level of fertility in Bolivia.[10]

NATIONAL POPULATION CENSUSES, 1831–1882

SECONDARY SOURCES

Dalencé, José M.
Bosquejo estadístico de Bolivia. Chuquianca, Imp. de Sucre, 1851. 391 pp.
New York Publ. Lib.

Population totals are given for the censuses of 1831, 1835, and 1846.

Pando Gutierrez, Jorge.
"Las actividades estadísticas de Bolivia." Pp. 33–57 in: Inter American Statistical Institute, *op. cit.* Washington, 1941. 842 pp. HA 175.S75

Total populations are given from the Peruvian Census of 1796 and the Bolivian Censuses of 1831, 1835, 1846, 1851, 1854, and 1882.

See also Tomo I, *Censo general de . . . 1900.*

CENSUS OF 1900

Oficina nacional de inmigración, estadística y propaganda geográfica.
Censo general de la población de la República de Bolivia según el empadronamiento de 1° de septiembre de 1900. La Paz, J. M. Gamarra, 1902–04. 2 vol.
HA 961 1900

Tomo I. Resultados generales. 29, xliv, 400 pp. [A general introduction gives the total and provincial population figures from the Census of 1831, 1835, 1845, 1854, and 1882 and describes the organization and procedures for the Census of 1900. Summary results of the 1900 census are then presented, giving for each province the population by urban-rural residence and by sex, for localities. A section for each province follows, giving the official census correspondence and the population for smaller localities by urban-rural residence and sex.]

Tomo II. Resultados definitivos. lxxxiii, 145 pp. [The first section, "Reseña geográfica y estadística de Bolivia," is followed by data on total population, age and sex, race, elementary education, marital status, religion, legal domicile, nationality, physical impairments, and occupations. Similar tables are then presented for each of the departments and the national territory.]

---

[7] Dirección general de estadística. *Demografía, 1940.* La Paz, Editorial Argote, 1942. 191 pp.
Govt. Publ. R. R.

[8] *Ibid.*, Nota de importancia, Preface.

[9] Dirección general de estadística. *Demografía, 1939.* La Paz, Fenix, 1940. 215 pp.

[10] Dirección general de estadística. *Demografía, 1940.* La Paz, Editorial Argote, 1942. 191 pp. For instance, on page 2, the number of births reported for the Province of Omasuyos, Department La Paz, is 2,726. Of these, 2,417 are male and 309 female. For the country, registered births were 27,119 male and 23,013 female, indicating a sex ratio of 118.

CITY POPULATION CENSUSES

LA PAZ

La Paz, Bolivia. Comisión central del censo.
Censo municipal de la ciudad de La Paz, 15 de junio de 1909. Clasificaciones estadísticas, precedidas de una reseña geográfica-descriptiva-histórica de la ciudad. La Paz, Tall. tip. lit. de José Miguel Gamarra, 1910. xiii, 80, ix pp.
HA 968.L2A5 1909

"Parte primera, Territorio," is a descriptive survey of the physical, geographic, biological, economic and historical aspects of the city. "Parte segunda, Censo de población," includes a retrospective summary of population estimates and censuses, 1586–1902. Municipal or national censuses are said to have been taken in 1675, 1796, 1831, 1845, 1886, and 1902. Data presented for 1909 include increase, density, nationality, race, sex, age, occupation, elementary education, religion, marital status, legal domicile, and physical impediments. An appendix presents tables on vital rates, 1890–1908; mortality by age, 1881–1909; mortality of adults and children, by sex, 1881–1909; diseases, 1900–1909; infirmities by age, 1900–1909.

Dirección general de estadística. Sección demográfica.
Censo demográfico de La Paz. La Paz, Typescript, 1943. Doc. Div.

Tables: Población de la ciudad el 15 de octubre de 1942 según el lugar de nacimiento. Población de la ciudad el 15 de octubre de 1942 según la raza de los habitantes.

CURRENT NATIONAL VITAL STATISTICS

(Including Population Estimates)

Dirección general de estadística.
Boletín anual, 1941. La Paz, 1942. 25 pp., graphs. Govt. Publ. R. R.

Births, marriages, and deaths are given by months for the capitals of the departments, based on provisional data from the Civil Register.

Boletín mensual de información estadística, enero a junio, 1937. Noviembre de 1937. Publicación oficial. La Paz, Imp. Atenea, 1937. 12 pp., tables.
Govt. Publ. R. R.

The first section, "Demografía," includes estimates of the total population and its racial composition for departments. Vital statistics are given for 1936 and the first half of 1937. Numbers only are given, without rates. The preface includes a discussion of the plans for a population census.

The *Boletín* continued to be issued to Sept., 1941.

Demografía, 1940. La Paz, Editorial Argote, 1942. 191 pp. Govt. Publ. R. R.

Vital statistics include births by legitimacy, race, and sex; literacy, dialect, age and occupation of parent; and various cross-classifications for departments and provinces. In addition, there are detailed statistics on marriages, general mortality, infant mortality and migration. Numbers only are given, without rates.

OTHER CURRENT NATIONAL POPULATION STATISTICS

Dirección nacional de estadística y estudios geográficos.
Anuario geográfico y estadístico de la República de Bolivia, 1919. La Paz, Imp. Artística, 1920. 712 pp., fold. maps. HA 961.A5

BRAZIL

Historical

The Abbé Correa da Serra made an estimate of the population of Brazil in 1776 on the basis of the data of the ecclesiastical registers. The only other known count of the population during the colonial period is that made by the Department of War in 1808.[1] Another attempt at a count was made in 1818, but estimates of the size of the population of Brazil in the first half of the nineteenth century continued to be conjectural estimates. A census was planned for 1851, but a multiplicity of factors prevented its realization. Official estimates were made in 1854 and 1867 on the basis of inquiries made by the presidents of the provinces. Decrees of 1870 and 1871 provided for the execution of a census of the Empire and a general statistical bureau was created. The Imperial census of 1872 constitutes the first approach to an actual enumeration of the population. This census, published in 23 volumes, presented detailed information not only on the distribution of the population but also on age, sex, nationality, marital status, color, and free or slave status.

Five national censuses have been taken during the history of the Republic, in 1872, 1890, 1900, 1920, and 1940. A census was planned for 1880, but the deterioration of the imperial statistical organization prevented its realization. The census of 1890 was taken under extraordinarily difficult circumstances; slavery had been abolished only 2 years before, and the federated Republic had replaced the Empire the previous year. The results of the census of 1900 were regarded as quite inaccurate at the time; those for the Federal District were not accepted, and a new census was taken in 1906. Data for the remainder of Brazil were examined and corrected for underenumeration and failure to enumerate, after which an estimate of the population of the states and municipalities was published. Political instability was responsible for the postponement and eventual cancelation of the census of 1910. Thus in 1920 Brazil had few reliable statistics which were not almost 50 years old. The comprehensive census of that year covered population, industry, and agriculture. However, there was considerable evidence that it represented a great overenumeration of the total population, and especially for that of certain areas. It remained the necessary base for the computation of vital rates and the analysis of Brazilian population problems for 20 years, since the outbreak of revolution necessitated the cancelation of the census planned for 1930.

The Brazilian Institute of Geography and Statistics, a cooperative association of the various State statistical departments, was organized in 1934 to coordinate the statistical services of the States and to determine standards and improve techniques for the reporting and analysis of social, economic, financial, and agricultural statistics. This agency was responsible for the preparation and execution of the census of 1940.

---

[1] Mortara, Giorgio. "A riddle resolved: Brazil's population." *Estadística, Journal of the Inter American Statistical Institute* 1 (1): 142–147. March 1943. For a more detailed history of the early estimates and censuses, see: "Resumo histórico dos inquieritos censitarios realizados no Brasil." Pp. 403–482 in: Diretoria geral de estatística. *Recenseamento do Brazil realizado em 1 de setembro de 1920.* Vol. 1. *Introducção.*

Also: Teixeira de Freitas, M. A. "As atividades estatísticas do Brasil." Pp. 71–112 in: Inter American Statistical Institute, *op. cit.*

Both a family and an individual schedule were used in the census of 1940. The individual schedule included 45 questions, covering name, sex, age, marital status, place of birth, education, employment status, and chronic infirmities.[2] Fertility questions included number of children ever born, number living, age of parents at birth of the first child, and number of children living at home. The question on color had as its purpose the determination of the components of Brazilian racial intermixture. Nativity questions included the place of birth of the individual and his parents, the date of entry into Brazil, the ability to speak Portuguese, and mother tongue. There were various questions on education. Occupational information included principal or current occupation and secondary occupation, and whether or not exercised for remuneration. The economic census consisted of five parts: agriculture, industry, commerce, transportation and communication, and services. The social census covered institutions and establishments serving various social needs.

Preliminary data only are available from this census. A volume has been issued, based on hand counts, giving the population of municipalities, and including a map of the distribution of the population.[3] However, the careful planning and administration of this census by the Institute of Geography and Statistics give promise of detailed analytical tabulations never before available for Brazil.

Vital statistics for the country, its regions, States, and principal cities are published in the *Anuário Brasileiro* of the *Instituto brasileiro de geografia e estatística.* The contents of the 1938 issue were republished, with additions, as a series of individual State yearbooks under the title, *Sinopse estatístico do estado. . . .* The *Boletim mensal do Serviçõ federal de bio-estatística* publishes current vital statistics for States. Monthly vital statistics for the city of Rio de Janeiro are often included in the *Revista brasileira de estatística.* In addition, the *Resumo da bio-estatística de cidades brasileiras,* issued biweekly, gives uncorrected vital statistics for a group of Brazilian cities, based on the civil register.

NATIONAL POPULATION CENSUSES

CENSUS OF 1872

Directoria geral de estatísca

Recenseamento de população do imperio de Brazil a que se procedeu no dia 1° de agosto de 1872. Quadros estatísticos. Rio de Janeiro, Leuzinger & filhos, 1873–1876. 23 vol. No copy located in U. S.

Copy in Bibliotheca Nacional, Rio de Janeiro. Title from: Rio de Janeiro. Bibliotheca nacional. Catalogo da exposicão de historia do Brazil a 2 de dezembro de 1881. Rio de Janeiro, Typ. de G. Leuzinger & filhos, 1881–1883. 329 pp.

Relatorio e trabalhos estatísticos apresentados ao illm. e exm. Sr. conselheiro Dr. João Alfredo Corrêa d'Oliveira, ministro e secretario de estado dos negocios do imperio pelo director geral interino Dr. Joaquim José de Compos da Costa de Medeiros e Albuquerque. Rio de Janeiro, Typographia e lithographia do movimiento, 1872. 320 pp., 89 fold. tables. HA 971.A2 1872

Relatorio e trabalhos estatísticos apresentados ao illm. e exm. Sr. conselheiro Dr. Carlos Leoncio de Carvalho, ministro e secretario de estado des negocios do imperio pelo director geral conselheiro Manuel Francisco Correia em 20 de novembro de 1878. Rio de Janeiro, Typographia nacional, 1879. 179 pp. HA 971.A2 1878

The data on education, physical defects, etc., are from the 1872 census.

[2] Carneiro Felippa, J. "A educação e a cultura no recenseamento geral de 1940." *Revista brasileira de estatística 2(7): 439–444.* July–Sept. 1941. The various parts of the census of 1940 are discussed, with emphasis on the questions concerning literacy and education.

[3] Instituto Brasileiro de geografia e estatística. Comissão censitaria nacional. *Recenseamento geral do Brasil realizado em 1° de setembro de 1940.* Sinopse preliminar dos resultados demográficos, segundo as unidades de federação e os municipios. Rio de Janeiro, 1941. 43 pp.

CENSUS OF 1890

Directoria geral de estatística.

Recenseamento geral da república dos Estados Unidos do Brazil em 31 de dezembro de 1890. Districto federal (cidade de Rio de Janeiro). . . . Rio de Janeiro, Typ. Leuzinger, 1895. xliii, 454 pp. [In Portuguese and French.] HA 971.A2 1890

Synopse do recenseamento de 31 de dezembro de 1890. Rio de Janeiro, Officina de estatística, 1898. x, 133 pp. [In Portuguese and French.] HA 971.A2 1890a

Sexo, raça e estado civil, nacionalidade, filiação, culto e analphabetismo da população recenseada em 31 de dezembro de 1890. Rio de Janeiro, Officina de estatística, 1898. 446 pp. [In Portuguese and French.] HA 971.A2 1890c

Idades da população recenseada em 31 de dezembro de 1890. Rio de Janeiro, Officina da estatística, 1901. 411 pp. [In Portuguese and French.] HA 971.A2 1890b

CENSUS OF 1900

Directoria geral de estatística.

Recenseamento de 31 de dezembro de 1900. Quadros de trabalho preliminar. Rio de Janeiro, Officina da estatística, 1900. 23 pp. HA 971.A2 1900

The mechanics of the census are described. No results are given.

Synopse do recenseamento de 31 de dezembro de 1900. Rio de Janeiro, Typographia da estatística, 1905. xv, 106 pp. [In Portuguese and French.] HA 971.A2 1900a

The results of this census for the Federal District were rejected and a new count made in 1906. See citation immediately below.

Recenseamento do Rio de Janeiro (Districto federal) realizado em 20 de setembro de 1906. Rio de Janeiro, Officina da estatística, 1907, xxvii, 399 pp. HA 988.R6A7 1906

This replaces the statistics for the Federal District secured in the national census of 1900.

CENSUS OF 1920

Numbered Census Volumes

Directoria geral de estatística.

Recenseamento do Brazil realizado em 1 de setembro de 1920. Rio de Janeiro, Typ. da estatística, 1922–1930. [L. C. set bound as 17 volumes.] HA 971.A2 1920

I. Introducção. Aspecto physico do Brazil. Geologia, flora e fauna. Evolução do povo Brazileiro. Historico dos inqueritos demographicos. 544 pp. [Résumé of population estimates, census history, and organization and methods of 1920 census included, pp. 403–544.]

I. Annexos. Decretos, instrucções, e modelos das cadernetas e dos questionarios para a execução do recenseamento. 1922. 160 pp. [Bound with Vol. I.]

II, 1a parte. População do Rio de Janeiro. Histórico da cidade e dos inqueritos censitarios. Crescimento, densidade e distribuição da população segundo o sexo, o estado civil, a nacionalidade, a idade, o gráo de instrucção, os defeitos physicos e as profissões. 1923. 648 pp.

II, 2a parte. Agricultura e industrias. Districto Federal. 1924. 192 pp. [Includes "Categoria e nacionalidade dos proprietarios."]

II, 3a parte. Estatística predíal e domiciliaria da cidade do Rio de Janeiro, Districto Federal. 1925. 548 pp., charts.

III, 1a parte. Agricultura. 1923. 512 pp., charts. [Includes "Categoria e nacionalidade dos proprietarios."]

III, 2a parte. Agricultura. 1925. 526 pp. [Continuation of material in 1a parte.]

IV, 1a parte. População. População do Brazil por estados, municipios e districtos, segundo o sexo, o estado civil e a nacionalidade. 1926. 883 pp.

IV, 2a parte. Tomos I, II. População. População do Brazil por estados e municipios, segundo o sexo, a idade e a nacionalidade. 1928. I, 795 pp. II, 868 pp.

IV, 3a parte. População. População do Brazil por estados e municipios, segundo os defeitos physicos, por idade, sexo, e nacionalidade. 1928. 265 pp.

IV, 4a parte. População. População do Brazil por estados, municipios e districtos, segundo o grão de instrucção, por idade, sexo e nacionalidade. 1929. 811 pp.

IV, 5a parte. Tomos I, II. População. População do Brazil por estados e municipios segundo o sexo, a nacionalidade, a idade e as profissões. 1930. I, 625 pp. II, 851 pp.

IV, 6a parte. Estatística predíal e domiciliaria do Brazil. 1930. 720 pp.

V, 1a parte. Industria. 1927. 526 pp.

V, 2a parte. Salarios. 1928. 520 pp.

V, 3a parte. Estatísticas complementares do censo economico. 1929. 210 pp.

OTHER PUBLICATIONS OF THE 1920 CENSUS [4]

Directoria geral estatística.

Recenseamento de 1920. Instrucções para a apuração do censo demographico. Rio de Janeiro, Typographia da estatística, 1922. 31 pp. Pan Am. Union

Recenseamento de 1920. Tabellas de conversão das principaes medidas agrarias usades no Brasil en unidades do systema metrico decimal. Rio de Janeiro, Typ. da estatística, 1921. ix, 104 pp. N. Y. Pub. Lib.

Synopse do recenseamento realizado em 1 de setembro de 1920. População do Brasil. Resumo do censo demographico por estados, capitaes e municipios. Confronto do numero de habitantes em 1920 com as populações recenseadas anteriormente. Rio de Janeiro, Typ. da estatística, 1922. 43 pp. HA 972.A5 1920a

Recenseamento do Brasil realizado em 1 de setembro de 1920. Synopse do censo da agricultura. Superficie territorial, area e valor dos immovies ruraes, categoria e nacionalidade dos proprietarios, systema de exploração. População pecuaria, producção agricola. Rio de Janeiro, Typ. da estatística, 1922. 90 pp. HD 1871.A5 1920

Synopse do recenseamento realizado em 1 de setembro de 1920. População pecuaria. Numero de animaes das varias especies de gado. Rio de Janeiro, Typ. da estatística, 1922. 51 pp. HD 9433. B8A5 1922

Recenseamento do Brazil realizado em 1 de setembro de 1920. Custo dos inqueritos demographico e economico. Rio de Janeiro, Typ. da estatística, 1923. vi, 9–36 pp. Pan Am. Union

Valor das terras no Brazil segundo o censo agricola realizado em 1 de setembro de 1920. Valeur des terres au Bresil d'après le récensement agricole realisé au 1er septembre 1920. Rio de Janeiro, Typ. da estatística, 1924. 51 pp. [In Portuguese and French.] HJ 4319.A28 1924

Synopse do recenseamento realizado em 1 de setembro de 1920. População do Brazil. Resumo do censo demographico segundo o sexo, o estado civil, e a nacionalidade dos habitantes recenseados nos estados e nas capitaes. Coefficientes da população do Brazil por sexo, estado civil, e nacionalidade em 1872, 1890, 1900, e 1920. Rio de Janeiro, Typ. da estatística, 1924. 62 pp. HA 972.A5 1920

Synopse do recenseamento realizado em 1 de setembro de 1920. População do Brazil. Resumo do censo demographico segundo o sexo, o idade, a nacionalidade e os defeitos physicos dos habitantes recenseados nos estados e nas capitaes. Coefficientes da população do Brazil por 1890, 1900 e 1920. Rio de Janeiro, Typ. da estatística. 118 pp. Pan Am. Union

Synopse do recenseamento realizado em 1 de setembro de 1920. População do Brazil. Resumo do censo demographico segundo o grão de instrucção, a idade, o sexo e a nacionalidade, nos estados e nas capitaes, coefficientes da população do Brazil, em 1872, 1890, 1900 e 1920, segundo o grão de instrucção, a idade, o sexo e nacionalidade. Rio de Janeiro, Typ. da estatística, 1925. 39 pp. Scripps Found.

Synopse do recenseamento realizado em 1 de setembro 1920. População do Brazil. Resumo do censo demographico segundo as profissões, a nacionalidade, o sexo e a idade dos habitantes recenseados nos estados e nas capitaes. Coefficientes da população do Brazil, segundo as profissões a nacionalidade e o sexo, em 1872, 1900, e 1920. População de facto e de dereito no Brazil e nos estados,

[4] The volumes listed here are either preliminary releases or summary analyses of the information published in more detail in the numbered census volumes.

em 1920. Resumo da estatística predial e domiciliaria nos estados e nas capitaes, em 1920. Densidade, predial e domiciliaria, em 1872, 1900 e 1920. Rio de Janeiro, Typ. da estatística, 1926. iv, 5, 210 pp.
HA 972.A5 1920b

Confirmação dos resultados do recenseamento demographico de 1920 e da estimative feita pela Directoria geral de estatística da população de 6 a 12 annos existente no districto federal em 21 de dezembre de 1926. Rio de Janeiro, Typ. de estatística, 1927. 15 pp. Scripps Found.

PLANS FOR CENSUS OF 1930

Diario official.

Decreto n. 18.994 de 19 de novembre de 1929. Da regulamento de decreto legislative n 5.730 de 15 de outubro de 1929, que autoriza a proceder ao recenseamento da república em setembro de 1930. Diario official, Nov. 26, 1929, p. 23,719–23,722. J 6.B8

Directoria geral de estatística.

Recenseamento de 1930. Tabellas de conversão das principaes medidas agrarias usadas no Brazil em unidades do systema metrico decimal. 2a edição. Rio de Janeiro, Typ. de estatística, 1930. xix, 136 pp. John Crerar Lib.

CENSUS OF 1940

Preliminary Volume

Instituto brasileiro de geografia e estatística. Comissão censitária nacional.

Recenseamento geral do Brasil, realizado em 1° de setembro de 1940. Sinopse preliminar dos resultados demográficos segundo as unidades da Federação e os municípios. Rio de Janeiro, 1941. 43 pp., map. Govt. Publ. R. R.

Part I gives the population of provinces and regions as of the censuses of 1872, 1890, 1900, 1920, and 1940. Part II consists of five general tables, giving distribution of municipalities by number, area, population, and density by provinces, and the distribution by area and population for the nation. Part III gives the number of districts and the area and population of municipalities.

Other

Instituto Brasileiro de geografia e estatística.

Decreto-lei 237, de 2 de fevereiro de 1938. Regula o inicio dos trabalhos do recenseamento geral da república em 1940 e da outras providencias. Decreto-lei 969, de dezembro de 1938. Dispõe sôbre os recenseamentos gerais do Brasil. Rio de Janeiro, Serviço gráfico do Instituto Brasileiro de geografia e estatística, 1939. 24 pp. Inter Am. Stat. Inst.

Instituto Brasileiro de geografia e estatística. Serviço nacional de recenseamento.

Coleção de decretos-leis sôbre o recenseamento geral da república em 1940. Rio de Janeiro, Serviço grafico do I. B. G. E., 1939. 47 pp. Inter Am. Stat. Inst.

The last decree included is of February 28, 1939.

Diario official.

Decreto-lei no. 2.141, de 15 de abril de 1940. Regulamenta a execução do recenseamento geral de 1940, nos têrmos di decreto-lei 969, de 21 de dezembro de 1938. *Diario official,* Rio de Janeiro, Secção 1, April 17, 1940, p. 6681–6689.
J 6.B8

Also published separately: *Instituto Brasileiro de geografia e estatística, Serviço nacional de recenseamento,* (1940?). 32 pp.

Finalidades do censo agrícola. Rio de Janeiro, 1939. 14 pp. Bur. of Cen.

Description of the objectives of the agricultural section of the general census planned for 1940.

Censo demográfico. Caderneta do agente recenseador. Recenseamento geral de 1940. Rio de Janeiro, 19—? HA 979. A3

A população do Brasil em 1920 e 1940. Mensario de statística, Órgão do sistema regional, Instituto Brasileiro de geografia e estatística, p. 3. July, 1940.
Govt. Publ. R. R.

Population by states; number of agricultural, commercial, industrial, transportation, and communication, and service establishments, total only.

Instituto Brasileiro de geografia e estatística.
"Resultados preliminares do censo demográfico de 1940, em confronto com os do censo de 1920 e com as estimativas oficiais." *Revista Brasileira de estatística 2(6): 313.* April–June, 1941. Govt. Publ. R. R.

RECENT PROVINCIAL CENSUSES

São Paulo. Commissão central do recenseamento.
Recenseamento demográphico, escolar e agricola-zoötechnico do estado de São Paulo, 20 de setembro de 1934. São Paulo, Imprenza official do estado, 1936. 15 pp. Not located

CURRENT NATIONAL VITAL STATISTICS

(Including Population Estimates)

Instituto Brasileiro de geografía e estatística.
Anuário Brasileiro. Rio de Janeiro, 19–. Ano I, 1908–1912. Ano II, 1930–1935. Ano III, 1935–1937. Ano IV, 1938. Ano V, 1939. HA 971.A32
The section on "Estado da população" in the 1939 issue presents summary population data for 1872, 1890, 1900, 1906, and 1920. The section on "Movimiento da população" includes the following four parts: I. "Registro civil." (Relative completeness of reporting by townships. Vital statistics for regions, states, and capital cities, 1937, 1938, and 1939, and for the country as a whole, by various characteristics, annually 1930–1938. Detailed vital statistics for the Federal District, 1937–1939.) II, III. "Imigração." (Temporary and permanent, by nationality, 1938 and 1939. Permanent, by social-economic characteristics, 1939.) IV. "Naturalizações." (By sex, nationality, and occupation, 1937, 1938, and 1939.)

Ministério da educação e saúde. Departamiento nacional de saúde. Serviço federal de bio-estatística.
Boletim mensal do Serviço federal de bio-estatística. II (3), Sept., 1942. Rio de Janeiro, Imprensa nacional, 1942. 26 pp. Bur. of Cen.
Separate tables for each state list live births, stillbirths, infant deaths, total deaths, and deaths by cause for June, 1942. The April issue (Vol. I, No. 10) gave summary data for the states, 1932–1941.

CURRENT PROVINCIAL VITAL STATISTICS

GENERAL

(State). Instituto Brasileiro de geografia e estatística. Departamento estadual de estatística.
"Synopse estatística do estado, No. 3." Separata, com acréscimos, do *Anuário estatístico do Brasil, Ano IV, 1938.* Bur. of Cen.
The yearbook, one for each state, consists primarily of reprinted information from the *Anuário Brasileiro,* Ano IV. Place and date of publication vary from state to state.

BAHIA

Bahia. Directoria de estatística.
Anuário estatístico de Bahia. Bahia, 1937. HA 988.B3A3
Ano XV, covering 1937, is the last issue located.

Bahia. Secretaria de educação e saúde. Departamento de saúde. Inspetoria de bio-estatística.
Boletim bio-estatístico. Bahia, 19—. Monthly. November, 1941, is the last issue available. Pan Am. San. Bur.

DISTRITO FEDERAL

Distrito Federal. Departamento de geografia estatística.
Anuário estatístico do Distrito Federal, anos VII e VIII. Rio de Janeiro, 1941. Not located
This issue contains data for the biennium 1939-1940.

MINAS GERAES

Minas Geraes. Secretario da agricultura, indústria, commércio a trabalho.
Atlas económico de Minas Geraes. 1938. 59 pp. Govt. Publ. R. R.
Information, maps, and diagrams on demographic factors are included.

Minas Geraes. Departamento estadual de estatística.
Boletim do Departamento estadual de estatística. Belo Horizonte, 1939–1940.
Govt. Publ. R. R.
Bi-monthly publication, numbered consecutively since first issue. Marriages, births, and deaths by cause, by age and sex, are given by districts for the first half of 1940 in No. 6. Sept.–Oct. 1940, pp. 37–41.

PERNAMBUCO

Pernambuco. Directoria geral de estatística.
Anuário estatístico. Recife, 1928–1936. HA 988.P4A3
Ano IX, covering 1935–1936, is the last issue available.

RIO GRANDE DO SUL

Rio Grande do Sul. Departamento estadual de estatística.
Anuário estatístico do Rio Grande do Sul: 1941. 1° volume: Situação fisica e demográfica. 2 volumes. Situação econômica. Porto Alegre, 1941. Bur. of Cen.

SANTA CATARINA

Santa Catarina. Departamento de saúde publica.
Sinopse de bio-estatística do estado, 1938.
Florianopolis, Imprenta oficial do Estado, 1939. Pan Am. San. Bur.

SÃO PAULO

São Paulo. Departamento de saúde da Secretaria de educação e saúde pública.
Resume mensal do movimento demógrafo-sanitario do estado de São Paulo por municípios. São Paulo, 1941. Pan Am. San. Bur

CURRENT CITY VITAL STATISTICS

Ministério da educação e saúde. Departamento nacional de saúde. Serviço federal de bio-estatística.
Resumo da bio-estatística de cidades brasileiras—semana de 18 a 24 de octubro de 1942. Rio de Janeiro, Imprensa nacional, 1942. Bur. of Cen.
This mimeographed weekly bulletin gives uncorrected figures from the civil register on live births, stillbirths, infant mortality, deaths, and causes of death for the following cities: Rio Branco, Manáus, Belém, São Louis, Teresina, Fortalesa, Natal, João Pessôa, Recife, Maceió, Aracaju, Salvador, Vitória, Niterói, D. Federal, São Paulo, Curitiba, Florianópolis, Porto Alegre, B. Horizonte, Goiânia, and Cuiabá.

BELO HORIZONTE

Belo Horizonte.
Anuário de estatística demografo—sanitaria de Belo Horizonte . . . Belo Horizonte, Minas Geraes, 1938. 180 pp. Pan Am. San. Bur.

CURITIBA

Curitiba. Agência municipal de estatística.
Boletim No. 1. Prefeitura municipal, 1941. Not located
This first bulletin of the reorganized statistical service of the capital of Parana covers population and vital statistics.

MARANHÃO

Maranhão. Directoria de estatística e publicidade.
Bio-estatística. Dez anos de dados-meteorologicos e demografo-sanitarios relativos a S. Luis, em tabelas e gráficos . . . Maranhão, Imprensa oficial, 1939. 75 pp.
Not located
Contains comparative data for various cities of Brazil, including vital statistics.

RECIFE

Recife. Departamento de saúde publica. Inspetoria de epidemiologia e bio-estatística.

Boletim mensal de bio-estatística, municipio de Recife, estado de Pernambuco. Recife, 19—. Pan Am. San. Bur.

Volume 44, No. 1, Jan., 1941, was the last issue located.

RIO DE JANEIRO

Instituto brasileiro de geografia e estatística.

"Nascimentos, casamentos e óbitos na cidade do Rio de Janeiro." *Revista brasileira de estatística 2(6): 387.* April–June, 1941. Govt. Publ. R. R.

The publication usually carries a table giving live births, stillbirths and deaths for the city of Rio de Janeiro by months.

SALVADOR

Salvador.

Revista de estatística e divulgação do municipio do Salvador. Ano I, No. 1. Salvador, Bahia, Divisão de estatística e divulgação do prefeitura, Jan., 1942. Not located

SÃO PAULO

São Paulo. Secretario dos negócios da educação e saúde publica. Departamento de saúde do estado. Secção de estatística sanitaria.

Boletim hebdomadário de estatística demografo-sanitario do municipio de São Paulo, 19—. Govt. Publ. R. R.

No issues after 1938 were located.

OTHER CURRENT NATIONAL POPULATION STATISTICS

COMPENDIA

Brazil. Departamento nacional do cafe.

Atlas estatístico do Brasil. Organizado por Carlos August Ribeiro Campos com a colaboraçao do Departamento nacional do cafe. Rio de Janeiro, 1941. 132 pp. Govt Pub. R. R.

This compendium includes a section on the demographic situation, pp. 16–21, giving vital statistics for 1936, immigration and emigration for 1937 by country of origin or destination, a résumé of population according to the first four censuses, and estimates for 1930 and 1938.

MIGRATION

Conselho de imigração e colonização.

"Primeiro ano de trabalhos de Conselho de imigraçáo e colonização." *Revista de imigração e colonização 1 (1): 5–19.* French summary, 19–22. Jan. 1940.

"Segundo ano de trabalhos de Conselho de imigração e colonização." *Revista de imigração e colonização 2 (1): 9–17.* Jan. 1941. Govt. Pub. R. R.

These are reports on the immigration policy of Brazil, made by the Secretary to the Council of Immigration and Colonization.

Ministério do trabalho, indústria e commércio. Departamento nacional de imigração.

"Imigrantes entrados no Brazil no período de 1884 a 1939." *Revista de imigração e colonização 1 (4): 617–644.* Oct. 1940. Govt. Pub. R. R.

A series of tables gives nationality, by individual years, 1884 through 1939, with a summary table by decades.

Ministério das relações exteriores.

Brasil 1940–41. Relação das condições geográficas, económicas e sociais. Rio de Janeiro, 1941. 481 pp. Govt. Pub. R. R.

State populations according to the first four censuses, p. 23. Chapter on immigration and colonization includes decennial immigration for 1884–1939, and immigration of Japanese decennially for 1904–1913.

Secretaria da agricultura, indústria e comércio.

Boletim do Serviço de imigração e colonização. São Paulo, 1937–19—. Irregular. Govt. Pub. R. R.

No. 4 contains detailed reports of immigration and emigration for the state of São Paulo for 1940.

STUDIES BY GIORGIO MORTARA, TECHNICAL CONSULTANT TO NATIONAL CENSUS COMMITTEE

"Estudos sôbre a utilização do censo demográfico para a reconstrução das estatísticas do movimento da população do Brazil." Series in: *Revista Brasileira de estatística, 1 (1 to 4) 1940 and 2 (5 to 6) 1941.* Govt. Publ. R. R.

I. "Estimativa do numero dos nascimentos." *Ibid.*, 1 (1): 7–16. Jan.–March 1940. [An analysis of methods of correcting errors in the reported numbers of births in an intercensal period, with applications to Brazilian statistics for the decade preceding the 1920 census.]

II. "Conjeturas sôbre os níveis da natalidade e da mortalidade no Brasil, no período 1870–1920." *Ibid.*, 1 (2): 229–247. April–June 1940. [Continuation, but with application to a fifty-year period.]

III. "Análise dos erros existentes nas distribuições por idade da população do Brasil, baseadas nos censos." *Ibid.*, 1 (3): 443–472. July–Sept. 1940. [Types of errors in the age distribution of the population, errors in the age group under five; errors by states, and for the Federal District; deficit in the first year of life and excess in the following years in the censuses of 1890 and 1920; anomalies of the census of 1872; errors in the rounding of age, with an analysis of such errors in the Federal District and in the city of São Paulo; other errors introduced by the attraction or repulsion exercised by certain figures; possibilities for the compensation of errors in the formation of age groups; other errors in age, i. e., understatement; and errors introduced in the execution of the census.]

IV. "Ensaio de ajustamento das tábuas de mortalidade Brasileiras calculadas por Bulhões Carvalho." *Ibid.*, 1 (4): 674–693. Oct.–Dec. 1940. [Revision of the life tables for single years of age for the Federal District and for the Brazilian capital cities taken collectively. Cities covered are: Belém, Fortaleza, Natal, Paráiba, Recife, Maceió, Salvador, Niterói, São Paulo, Curitaba, Florianópolis, Pôrto Alegre, Belo Horizonte.]

V. "Rétificação da distribuição por idade da população natural do Brasil, constante dos censos; cálculo dos óbitos, dos nascimentos e das variacões dessa população no período 1870–1920." *Ibid.*, 2 (5): 39–89. Jan.–March 1941. [Corrected data are given from 1870–1871 to 1914–1920, with calculations of age distributions conforming to the hypotheses adopted with reference to natality and mortality. Mortality rates are computed for the native population of Brazil by quinquennial age groups, and the mortality of the population of alien origin added.]

VI. "Sinopse da dinâmica da população do Brasil nos últimos cem anos." *Ibid.*, 2 (6): 267–276. April–June 1941. [Corrected estimates at quinquennial periods, 1870–1920; average population, native, alien, and total, for the same periods; death rates of natives and aliens; rates of births, deaths and natural increase, and supplementary calculations on the hypothesis of errors in the 1920 census.]

VII. "Tábuas de mortalidade e de sobrevivência para os períodos 1870–1890 e 1890–1920. Cálculo, exame e comparações internacionais." *Ibid.* 2 (7): 493–538. July–Sept., 1941. [A description of data and techniques precedes a comparative analysis of the age distribution of deaths; mortality in infancy, adolescence, and the central life span; and the average duration of economically productive life. International comparisons are included.]

Os fatores demográficos do crescimento das populações americanas nos ultimos cem anos. Rio de Janeiro, 1940. Inter Am. Stat. Inst.

Components in the population growth of individual countries of the Americas are computed, with comparisons by language groups. Brazil and Argentina are discussed in detail.

I. *Nota sôbre a população do origem ou de lingua Alemã no Brasil.* Rio de Janeiro, Dec. 28, 1941. 2 pp. Bur. of Cen.

II. *Dados e cálculos sôbre a imigração Alemã no Brasil.* Rio de Janeiro, Dec. 28, 1941. 4 pp. text, 4 pp. tables.

III. *Dados e cálculos sôbre a imigração Austríaca no Brasil.* Rio de Janeiro, Dec. 30, 1941. 2 pp. text, 4 pp. tables.

IV. *Dados e cálculos sôbre a imigração Italiana no Brasil.* Rio de Janeiro, Jan. 15, 1942. 10 pp.

V. *Dados e cálculos sôbre as imigrações Húngara, Rumena, Bulgara e Finlandesa no Brasil.* Rio de Janeiro, Jan. 15, 1942. 3 pp. text, 2 pp. tables.

VI. *Dados e cálculos sôbre a imigração Japonese no Brasil.* Rio de Janeiro, Dec. 1941. 8 pp.

Sinopse dos números de naturais de alguns países estrangeiros atualmente existentes no Brasil. Rio de Janeiro, Jan. 12, 1942. 1 p.

Contribuição ao estudo da assimilação matrimonial e reprodutiva dos diferentes grupos estrangeiros na população do Brazil. Rio de Janeiro, 1942. 28 pp.
Inter Am. Stat. Inst.

Based upon data for the city of São Paulo for the biennia 1920–21 and 1938–39.

A população do Brasil, por regiões fisiográficas, conforme os resultados provisórios do censo de 1940 e a nova divisão territorial adotada pelo Instituto Brasileiro de geografia e estatística. Rio de Janeiro, 1942. 3 pp. Inter Am. Stat. Inst.

Estimativa do número dos centenários no Brasil em 1940 e análise comparativa international da apuração dos centenários pelos recenceamentos. Rio de Janeiro, 1942. 25 pp. Inter Am. Stat. Inst.

Tábuas de mortalidade e de sobrevivência para a capital federal e a capital de São Paulo, anos 1920–21 e 1939–1940. Rio de Janeiro, 1942. 4 pp.
Inter Am. Stat. Inst.

Expectation of life is given for years 0–5, and for decennial years.

MISCELLANEOUS

Instituto Brasileiro de geografia e estatística.

Legislação organica do sistema estatístico-geográfico Brasileiro, 1934–1939. Vol. I: Organização nacional. Rio de Janeiro, Serviço gráfico do I. B. G. E., 1940. 104 pp. Inter Am. Stat. Inst.

Decrees and laws establishing and implementing the Brazilian Institute of Geography and Statistics.

Instituto Brasileiro de geografia e estatística.

Divisão territorial dos Estados Unidos do Brasil. Quadro territorial—administrativo e judiciario—das unidades de Federação, fixado para o quinquênio de 1939–1943, em virtude de lei orgânica nacional no. 311, de 2 de março de 1938. Rio de Janeiro, Serviço gráfico do Instituto Brasileira de geografia e estatística, 1940. 451 pp. Inter Am. Stat. Inst.

The first part consists of two tables for each of the Brazilian states. Table 1 gives the names of the judicial, administrative, and joint administrative-judicial areas, respectively. Table 2 lists districts which through recent laws have changes in area or classification, with a description of the change.

The second part consists of two sections. Section I lists changes in place names. Section II lists cities and towns of circumscribed area, giving the political unit to which they pertain.

The Appendix (pp. 365–451) cites the laws authorizing the changes.

CHILE

Historical

There were three enumerations of the population of Chile in the eighteenth century. A nation-wide census was ordered in 1811, but only the Province of Concepción actually completed its enumeration. A general census was attempted in 1813–14, but the few surviving fragments are insufficient for any evaluation of its extent or validity. The first national census was taken in 1831 and 1835. Counts were made of the population in Maule, Concepción, Valdivia, Chilóe, and Santiago in 1831, while the remaining provinces were covered in the year 1835.[1] A second census, taken in 1843, was completed within a year. The accuracy of both the 1831–35 and 1843 counts is quite questionable, since techniques were simple and the population widely dispersed and uncooperative.[2] No publications of either census have been located.

In 1853 a law was passed providing for the taking of a census in the following year and at 10-year intervals thereafter. The census authorized for 1854 was actually taken, and the results published in some detail by the *Oficina central de estadística* in 1858. Subsequent censuses were taken in 1865, 1875, 1885, 1895, 1907, 1920, 1930, and 1940. The final volumes of the 1940 census are not yet available, although preliminary results have been published in *Estadística Chilena,* the monthly bulletin of the *Dirección general de estadística.*

Vital statistics have been recorded in Chile since 1848, although a civil registration system was not established until 1885.[3] A centralized card reporting system has been used since 1912; current vital statistics are now published in the monthly bulletin, *Estadística Chilena.* The December issue released in March of the following year, contains a yearly summary. Current population and vital statistics are also made available in the *Anuario estadístico de Chile.* This publication consists of seven volumes, the first, *Demografía y asistencia social,* contains about 100 pages of population statistics.

NATIONAL POPULATION CENSUSES

CENSUS OF 1854

Oficina central de estadística.
Censo jeneral de la República de Chile levantado en abril de 1854. Santiago, Imprenta del ferrocarril, 1858. 9 pp. text, 43 pp. tables. HA 991. A2 1854

CENSUS OF 1865

Oficina central de estadística.
Censo jeneral de la República de Chile, levantado el 19 de abril de 1865. Santiago, Imprenta nacional, 1866. xxvii, 396 pp. HA 991. A2 1865

---

[1] Oficina central de estadística. *La población de Chile,* por Francisco de Béze, director. Santiago, Chile, Impr. Bellavista, 1911. 50 pp. Summary statistics on population growth from 1700 to the time of writing are presented, based on early estimates and the later census counts.

[2] Inter American Statistical Institute. *Op. cit.,* pp. 756–757. For a concise statement of the census history of Chile, *see:* Vergara, Roberto. "Los censos de población en Chile." pp. 95–108 in: *Eighth American Scientific Congress. Proceedings, Vol. VIII.* Washington, Dept. of State, 1942. 365 pp.

[3] Vergara, Roberto. "Las actividades estadisticas de Chile." pp. 172–198 in: Inter American Statistical Institute, *op. cit.*

CENSUS OF 1875

Oficina central de estadística.
Quinto censo jeneral de la población de Chile levantado el 19 de abril de 1875 i compilado por la Oficina central de estadística en Santiago. Valparaíso, Imprenta del Mercurio, 1876. lviii, 674 pp. HA 991. A2 1875a

CENSUS OF 1885

Oficina central de estadística.
Sesto censo jeneral de la población de Chile levantado el 26 de noviembre de 1885 y compilado por la Oficina central de estadística en Santiago. Valparaíso, Imprenta de "La Patria", 1889–1890. 2 vol. HA 991. A2 1885

Censo de población de subdelegaciones i departamentos de la República de Chile. Cuadro de departamentos y subdelegaciones existentes en 26 de septiembre de 1887, y su correspondiente población con arreglo al censo de 1885. Santiago, Imprenta nacional, 1887. 28 pp. HA 991. A4 1887

CENSUS OF 1895

Oficina central de estadística.
Sétimo censo jeneral de la población de Chile levantado el 28 de noviembre de 1895 y compilado por la Oficina central de estadística. Valparaíso, Imprenta Guillermo Helfmann, 1900. Pan Am. Union

CENSUS OF 1907

Comisíon central del censo.
Censo de la República de Chile levantado el 28 de noviembre de 1907. Santiago, Sociedad "Imprenta y litografía Universo", 1908. xlvii, 1320 pp. HA 991. A2 1907

CENSUS OF 1920

Dirección general de estadística.
Censo de población de la República de Chile levantado el 15 de diciembre de 1920. Santiago, Soc. imp. y litografía Universo, 1925. xxxi, 610 pp. HA 991. A4 1920

CENSUS OF 1930

Laws, statutes, etc.
Decreto y reglamento relativo al décimo censo nacional de la población y ley 4541 de 25 de enero de 1929. Seguros y impuestos, Santiago, June, 1930, pp. 1433–1438. Pan Am. Union

Dirección general de estadística.
Décimo censo nacional de la población, 27 de noviembre de 1930. Folletos no. 1–5, 7. Santiago, Imprenta Lagunas & Quevedo, ltda, 1930. Pan Am. Union

Numero 1. Instrucciones preliminares que la comisión central imparte a las comisiones departamentales y comunales. Enero de 1930.

Numero 2. Instrucciones sobre la división de la comuna en zonas de empadronamiento. Febrero de 1930.

Numero 3. Historia y fines de los censos: los censos de Chile, sus errores; determinación de la población entre censo y censo; preparación y realización de un censo; el censo chileno de 1930. Por el ingeniero Don Roberto Vergara. Febrero de 1930.

Numero 4. Decreto y reglamento relativo al décimo censo nacional de la población y ley 4541 de 25 de enero de 1929. Marzo de 1930.

Numero 5. Resumen de la labor preparatoria del censo efectuada en los 5 primeros meses de 1930. Junio de 1930.

Numero 6. Not located.

Numero 7. Instrucciones para los empadronadores. Septiembre de 1930.

Dirección general de estadística. Comisión central del censo.
Resultados del X censo de la población efectuado el 27 de noviembre de 1930 y estadísticas comparativas con censos anteriores. Santiago, Imp. Universo, 1931–1935. 3 vol. HA 991. A4 1930

Volumen I. Estadísticas comparativas con censos anteriores. 1931. 298 pp. [The preparations for the census are discussed and the instructions to the enumerators and local officials are reproduced. The main body of the volume reports data on the distribution of the population according to civil divisions, size classes, urban-rural distribution, and sex.]

Volumen II. Edad, estado civil, nacionalidad, religión e instrucción. 1933. xi, 512 pp. [The factors listed are reported in detail for provinces and minor civil divisions.]

Volumen III. Ocupaciones. 1935. 171 pp. [The population is distributed according to occupation, nationality, and age groups under 15, 15–19, 20, and over.]

CENSUS OF 1940

Dirección general de estadística. Comisión directiva del censo.
Reglamento del XI censo de población. Santiago, Imp. y lito. Universo, 1940. 3 vol. No. 1, 14 pp.; No. 2, 26 pp.; No. 3, 32 pp. Pan Am. Union.
"Folleto 1, Reglamento del XI censo de población (dtos. 136 y 483)," is in the library of the Pan American Union. The other pamphlets were not located.

Dirección general de estadística. Sección geografía administrativa.
División administrativa por provincias, departamentos y comunas, con las circumscripciones del registro civil. Población según el censo de 1940. Santiago, 1941. Fold. table. Govt. Publ. R. R.
This table reports the population of Chile at the time of the 1940 census.

Dirección general de estadística.
Censo. Resultados generales del XI censo de población, por provincias, departamentos, comunas y distritos. Estadística Chilena 14 (12): 659–673. Dec. 1941.
Censo. Número de viviendas por categoria, provincias, departamentos, comunas y distritos . . . Provincia de Coquimbo. Ibid., 15 (9): 408–420. Sept., 1942. HA 993.A3

OTHER NATIONAL CENSUSES

AGRICULTURE, 1929–1930

Dirección general de estadística.
Censo agropecuario, 1929–1930. Santiago, 1938. 1 vol. Pan Am. Union.
Detailed information on acreage, yield, livestock, and so forth; does not include information on agricultural labor force.

EDUCATION, 1933

Dirección general de estadística.
Censo de educación. Año 1933. Santiago, Imp. Universo, 1934. 84 pp. L 301.B5 1933
Includes age of students and marital status and training of teachers for all types of schools.

INDUSTRY AND COMMERCE, 1928, 1937

Dirección general de estadística.
Censo de la industria manufacturera y del comercio de 1928. Santiago, Imprenta Universo, 1929. 119 pp. HA 191.A45 1928
There are two sections: A. "Industria manufacturera," and B. "Comercio interior." Both include nationality of proprietors and classification of personnel by type of employment (by month or day), sex, locality, and industry.
Censo industrial y comercial, año 1937. Santiago, Imprenta y litografía Universo s. a., 1939. 241 pp. HC 191.A45 1937
Information is similar to that in the census of 1928, including nationality of proprietors and classification of personnel by type of employment (by month or day), sex, locality, and industry.

CURRENT NATIONAL VITAL STATISTICS

(Including Population Estimates)

Dirección general de estadística.
Estadística Chilena. Santiago, Imprenta y litografía Universo s. a., Vol. 15, No. 1–2. Jan.-Feb., 1942. HA 993.A3
The section, "Demografía," regularly includes the following data: Marriages; live births and stillbirths by legitimacy status and sex; deaths by age and sex, by provinces and cities; and deaths by cause. The December issue of each year summarizes vital statistics for the year. Annual marriage, birth and death rates for 1918 to 1941 were included in the issue for Dec., 1941.
Estadística anual. Volumen I: Demografía y asistencia social. Santiago, Soc. imprenta y litografía Universo, 1940. HA 991.B2

Vital statistics for 1939 are presented in greater detail than in the monthly *Estadística Chilena.* Age, sex, occupation, nationality, and other classifications are used, including geographical break-down by zones.

Servicio nacional de salubridad.

Anexo estadístico a la memoria de Servicio nacional de salubridad correspondiente al año 1939. Santiago, 1941. 113 pp. Govt. Publ. R. R.

Tables 32–97 give vital statistics for 1939, with comparative data for the period 1932–1939. The final section, pp. 93–113, consists of 105 graphs of population and vital statistics for various periods through 1939. Morbidity and mortality for specific diseases are included.

Ministerio de salubridad, previsión y asistencia social. Departamento de previsión social.

La seguridad social, por el Dr. Julio Bustos A., jefe del departamento de previsión social del Ministerio de salubridad, previsión y asistencia social. Estudio de la previsión social en Chile, sus resultados después de diez años de aplicación y las reformas que deben introducirse. Trabajo efectuado con la colaboración del personal del Departamento de previsión social. Santiago, 1936. 266 pp. HD 7156.A4 1936

The introductory section contains general mortality statistics for Chile, and an analysis of the relative importance of the different causes of death.

Ministerio de salubridad.

La realidad médico-social Chilena. Síntesis. Santiago, Imprenta Lathrop, 1939. 216 pp. RA 465.A52 1939

Part I, "Algunos antecedentes geográficos y demográficos," summarizes trends in population increase, age composition, fertility and mortality for various periods through 1938, in some cases from 1848. Part II, "Condiciones de vida de las clases trabajadoras," is concerned with economic conditions and the analysis of mortality for specific diseases in the light of those conditions.

OTHER CURRENT OFFICIAL POPULATION STATISTICS

Dirección general de estadística.

Sinopsis geográfico-estadística de la República de Chile. 1933. Santiago, 1933. 1 vol. HA 992.A33

Population data for the census years from 1854 through 1930, and vital statistics from 1848 through 1932 are included.

Madrones Restat, Jorge.

Sobre el cálculo de la duración media de la vida en Chile. Santiago, Impreso en los talleres gráficos de la editorial "Cultura", 1940. 10 pp. (From: Boletín de salubridad.) Pan Am. Union

1930 census data and vital rates for 1848–1930 are the basis for the calculations.

COLOMBIA

Historical[1]

The Colombian Constitution of 1821 decreed that representation of the individual districts of the country should be according to population. A series of censuses was taken under this authorization, in 1825 for Greater Colombia, and in 1835, 1843, and 1851 for Nueva Granada. These censuses were taken by enumerators in the districts, who compiled the returns and sent only summaries to the central government. Many factors influenced the accuracy of this census. The use of the results for determining representation tended to maximize returns, while the usual fears of taxes and military service led to under-enumeration.

A census law of 1858 transferred the function of census taking to the Federal States, under an administrative system which tended to produce a series of individual State censuses rather than a Federal census. This defect is particularly apparent in the census of 1864, which was published only as a count of the number of inhabitants. Data on age and occupational composition were published for the first time in the census of 1870, which was the last general census before the civil war. The new census law of 1904 provided that censuses were to be taken by the municipalities, but this method proved unsatisfactory. The census of 1912 was taken according to the European tradition, each individual filling out his own census questionnaire.

The census of 1918 was also conducted by the distribution of individual questionnaires and counting by individual districts. It included age distributions by quinquennial groups, as well as information on education and occupations. A census of 1928 was carried out by the *Contraloría general*, but it was so incomplete that the Congress refused to accept it.

In preparation for the 1938 census, a *Dirección general de los censos* was set up independently of the political administration. Careful plans were made for the organization and execution of the census. A census of housing, later published in 1939, was carried out on April 20, 1938, preliminary to the total census. The questionnaire of the main census covered name, position of the head of the family, sex, age, marital status, degree of education, nationality, occupation, religion, and physical and mental defects.

Vital statistics have a long history in Colombia, but they remain incomplete. The ecclesiastical register recorded the marriages, baptisms, and funerals of Catholics. A civil registration office was set up in 1873, but was not used completely. A law of 1914 provided for a third statistical register of marriages, births, and deaths, but deaths alone tended to be recorded because of the legal requirement of burial permits. The national reports of the *Contraloría general* have been based on combinations of information from these various sources. A law of 1938, providing for a compulsory system of civil registration, offers possibilities for the development of more adequate vital statistics.

---

[1] Hermberg, Paul. "Las actividades estadísticas de Colombia." pp. 208–245 in: Inter American Statistical Institute, *op. cit.*

NATIONAL POPULATION CENSUSES

CENSUS OF 1827

Republic of New Granada. Secretario de estado.
Memoria. Esposición que el Secretario de estado del despacho . . . al Congreso de 1827. Bogotá, Imprenta de Pedro Cubides, 1827. 36 pp.
J 215.R2 1827

Two in-folded tables give summary results of the census of 1825. The data include Ecuador, which was not separated from Colombia until 1832.

CENSUS OF 1835

Republic of New Granada.
Censo de población de la República de la Nueva Granada, levantado con arreglo a las disposiciones de la lei de 2 de junio de 1834 en los meses de enero, febrero, i marzo del año de 1835 en las diferentes provincias que comprende su territorio. Gaceta de la Nueva Granada, trim. 16, no. 211. Oct. 11, 1835. Unnumbered. J 6.C6

Resumen del censo jeneral de población de la República de la Nueva Granada, levantado con arreglo a las disposiciones de la lei de 2 de junio de 1834 en los meses de enero, febrero i marzo del año de 1835 en las diferentes provincias que comprende su territorio i distribuído por provincias, sexos, edades i clases. Gaceta de la Nueva Granada, trim. 17, No. 234. March 20, 1836. Unnumbered. J 6.C6

CENSUS OF 1843

Republic of New Granada. Departamento del interior i relaciones esteriores.
Decreto del poder ejecutivo sobre formación del censo de población de la República. [Bogotá]. 7 pp. unnumbered. 1842. HA 37.C72

Republic of New Granada. Secretario de estado.
Esposición que el Secretario de estado en el despacho de lo interior dirije al Congreso constitucional de 1844. Bogotá, Imprenta de Cualla, 1844. Various pagings.
J 216.R2 1844

Tables 1–17 present data from the Census of 1843.

Republic of New Granada.
Resumen del censo jeneral de la población de la Nueva Granada, distribuído por provincias, cantones, i distritos parroquiales, con espresión del número de electores que a cada cantón i provincia corresponde, con arreglo al artículo 17 de la constitución. Gaceta de la Nueva Granada, trim. 50, no. 661, pp. 2–5. Jan. 7, 1844.
J 6.C6

Republic of New Granada, Departamento de relaciones esteriores.
Estadística jeneral de la Nueva Granada, que conforme al decreto ejecutivo de 18 de diciembre de 1846. Parte primera: Población e instituciones. Bogotá, Impr. de J. A. Cualla, 1848. 231 pp. HA 1014.A3 1848
Census of 1843, pp. 29–170

CENSUS OF 1851

Republic of New Granada. Secretario de estado.
Informe del Secretario de estado del despacho de gobierno de la Nueva Granada al Congreso constitucional de 1852. Bogotá, Imprenta del Neo-Granadino, 1852. 44 pp., excl. tables at end. J 216.R2
Census of 1851, tables 1–13.

CENSUS OF 1864

United States of Colombia. Secretario de relaciones esteriores.
Esposición del Secretario de lo interior y relaciones esteriores de los Estados Unidos de Colombia al Congreso de 1866. Bogotá, 1866. 108 pp. JX 552.A2 1866
Census of 1864, pp. 52–55. The only data published for this census are the total populations of the federal states.

Memoria de la Secretario de lo interior y relaciones esteriores al señor presidente de los Estados Unidos de Colombia, 1867. Bogotá, Imprenta de Echeverría Hermanos, 1867. 23 pp., excl. tables. JX 552 A2 1867
Census of 1864, pp. 21–22 and Table A.

CENSUS OF 1871 (1870?)

United States of Colombia, Secretario de relaciones esteriores.
Memoria del Secretario de lo interior i relaciones esteriores de los Estados Unidos de Colombia para el Congreso de 1875. Bogotá, Imprenta de Medardo Rivas, 1875. JX 552.A2 1875
Census of 1871 (1870?), pp. 120–122.

CENSUS REGULATIONS, 1899

Republic of Colombia. Dirección general de estadística.
Reglamento para la formación del censo y movimiento de la población. Bogotá, Imprenta nacional, 1899. 114 pp. HA 37.C8 1899

CENSUS OF 1905

Republic of Colombia.
Censo de población, 1905. Diario oficial, Vol. 53, No. 16028, pp. 489–496. Feb. 24, 1917. J 6.C6

CENSUS OF 1912

Republic of Colombia. Junta central del censo nacional.
Compilación de las disposiciones legales, ejecutivas y administrativas, que sirven de guía á todas las personas que desempeñen cargos en el levantamiento del censo. Ed. oficial. Bogotá, Imprenta nacional, 1911. 28 pp. HA 1012.A3 1911

Republic of Colombia. Ministerio de gobierno.
Censo general de la República de Colombia, levantado el 5 de marzo de 1912. Población–división política–municipios que componen la República–cuadros de clasificaciones–cuadros gráficos–reseñas estadísticas de los departamentos, con datos históricos y planos de sus capitales. Gobiernos civil y eclesiástico del país, etc. Bogotá, Imprenta nacional, 1912. 336 pp. HA 1012.A3 1912

CENSUS OF 1918

Republic of Colombia. Dirección general de estadística.
Censo de población de la República de Colombia levantado el 14 de octubre de 1918 y aprobado el 19 de septiembre de 1921 por la Ley 8ª del mismo año. Director general estadística, Alberto Schlesinger. Bogotá, Imprenta nacional, 1924. 448 pp. HA 1012.A3 1924

CENSUS OF 1928

Republic of Colombia. Contraloría general. Dirección del censo.
Memoria y cuadros del censo de 1928. Bogotá, Editorial Librería Nueva, 1930. Not located

CENSUS OF 1938

Republic of Colombia. Dirección general de los censos.
Decretos y bases de organización para la ejecución de los censos. Bogotá, Editorial El Gráficos, 1938. 60 pp. HA 1012.A32
Instructions to enumerators, with data on and organization charts of the 1938 census.

Informe que la Contraloría general de la República rende al Sr. Ministro de Gobierno y a las honorables Cámaras sobre el levantamiento del censo civil de 1938. *Anales de economía y estadística* 2 (4): 3–45. Aug., 1939. Govt. Publ. R. R.
There is a general introductory statement on the organization and results of the civil census of 1938. Part I, Antecedents, gives a brief history of the census in Colombia and of the preparations for this census. Part II, Legal basis, reproduces the various decrees. Part III, Technical basis, lists the data requested and defines terms. Part IV, General organization, discusses costs and describes the organization of the census. Part V, Results, summarizes basic data on trends and distribution.

Censo general de la población, 5 de julio de 1938. Ordenado por la ley, 67 de 1917. Bogotá, Imprenta nacional, 1940–1942. HA 1012.A3 1938

Tomo I. Departamento de Antioquia, 1940. 435 pp.
Tomo II. Departamento de Atlántico, 1940. 139 pp.
Tomo III. Bolívar. Not located.
Tomo IV. Departamento de Boyacá. 1940. 569 pp.
Tomo V. Departamento de Caldas. 1941. 237 pp.
Tomo VI. Departamento de Cauca. 1940. 193 pp.
Tomo VII. Departamento de Cundinamarca, 1941. 519 pp.
Tomo VIII. Huila. Not located.
Tomo IX. Departamento del Magdalena. 189 pp.
Tomo X. Departamento de Nariño. 255 pp.
Tomo XI. Departamento Norte de Santander. 1941. 191 pp.
Tomo XII. Departamento de Santander. xxxvi, 359 pp.
Tomo XIII. Departamento de Tolima. xxxi, 217 pp.
Tomo XIV. Departamento del Valle de Cauca. xxxi, 207 pp.
Tomo XV. Intendencias y comisarías. xxxi, 262 pp.
Tomo XVI. Resumen. xx, 195 pp.

The contents of each of the provincial volumes are divided into two parts—Censo de población and Censo de edificios. Population data include density, natural increase, rural-urban distribution, distributions by minor civil divisions; sex and marital status, literacy by rural-urban residence, age and sex; and nationality and occupation. The final summary volume contains similar data for the entire country.

OTHER NATIONAL CENSUSES

Dirección general de los censos.

Primer censo nacional de edificios. Efectuado el 20 de abril de 1938. Bogotá, Colombia, Imprenta nacional, 1939. 393 pp. TH 45.A5 1938

CURRENT NATIONAL VITAL STATISTICS

(Including Population Estimates)

Dirección general de estadística.

Anuario general de estadística. 1940. Bogotá, Imprenta nacional, 1941. HA 1011.A16

Vital statistics include marriages by age and marital status for departments and by nationality; births, 1931–1940, by legitimacy status, by nationality of parents and of mothers; deaths, 1931–40; deaths in 1940 by age, sex, and civil status; infant mortality by sex, and legitimacy status.

Migration statistics for 1935–1940 include information on sex, marital status, occupation and nationality. Population estimates are also included.

Estadísticas, demográfica y nosológica, 1936. Publicación dirigida por Miguel Angel Fscobar, director de la Sección de demografía de la Dirección nacional de estadística. Editada por cuenta del Ministerio de trabajo higiene y previsión social. Bogotá, Imprenta nacional, 1938. 63 pp. HA 1012.A4 1936

Vital statistics include marriages by marital status, age and nationality; births for departments and by nationality of parents; deaths by month, for departments, and by age, according to presence of medical assistance. Detailed statistics on the causes of death are presented.

Síntesis estadística de Colombia, 1941. Bogotá, Imprenta nacional, 1942. 155 pp. Govt. Publ. R. R.

The population section, pp. 22–33, includes a résumé of population growth, 1770–1938, summary data from the 1938 census, and estimated populations as of Dec. 31, 1941. Vital statistics are given for 1938–1941 inclusive. Data on net migratory change cover the years 1939, 1940, and 1941.

OTHER CURRENT NATIONAL POPULATION STATISTICS

Dirección general de estadística.

Estudio histórico analítico de la población colombiana en 170 años. Juan de D. Higuita. Anales de economía y estadística. Suplemento al número 2°, Tomo III. April 25, 1940. Bogotá, 1940. 113 pp. HA 3567.N5

This study, made under the auspices of the Director of Statistics, covers the census history of Colombia in four periods: colonial period to independence, about 1825; the national republic, to 1870; the period of civil strife, to 1905; and the thirty years of peace, 1905 to the present. The first part of the study characterizes the political and social life of these four periods and examines critically the censuses taken, including that of 1938. The second part is a mathematical analysis of population increase, with special emphasis on the validity of the logistic law of population growth. Population is projected to 1948.

"La tabla de mortalidad en Colombia." *Anales de economía y estadística 6 (3): II.* Feb., 1943. HA 3567.N5

The *Dirección nacional de estadística* announces that a life table for Colombia is to be published in the *Anales* in the near future. This table, the first official life table for the nation, was prepared by Jorge Rodríguez.

Sociedad colombiana de demografía.

La demografía colombiana. Año 1, Número 1. Jan., 1936. 53 pp. HB 881 A1 D5

Contents: Statutes of the Colombian Society of Demography; biometrics: the census of population; the population of Colombia; vital statistics in the Department of Bolívar; vital statistics in Bogotá; infant mortality in Bogotá; social statistics; and the Law of 1935 on the civil census of the Dominican Republic.

This publication was issued under the auspices of the *Sección de estadística, Contraloría general de la República.* Vol. 1, No. 1 appears to have been the only issue.

COSTA RICA

Historical

Many estimates of the size and racial composition of the population of Costa Rica have been made since the first Spanish contact with the area in 1502. Some of these were official estimates, based on special surveys or parish registers, while others were made by individuals or ecclesiastical institutions. B. A. Thiel, in his *Monografía de la población de Costa Rica en el siglo XIX*, traces the growth of the population from 1564 to 1801 on the basis of old parish records, the reports of various governors, and an informal census of 1771–78.[1]

Various reports on the population of Costa Rica cite data reputedly based on censuses in the years 1824, 1836, 1844, 1864, 1883, 1888, 1892, and 1927.[2] The present statistical office in Costa Rica admits only four of these as national censuses, those of 1844, 1864, 1892, and 1927. The status of the 1844 data is questionable; it appears to represent a compilation of data from parish registers. The report on the census of 1892 contains references to a census of 1883, but the *Departamento de biodemografía* was unable to locate it.[3] It appears probable that the information published for the other years consists of estimates of varying degrees of validity. For instance, in 1888 the *Dirección general de estadística* published a report entitled: *Población de la República de Costa Rica el 31 de diciembre de 1888.* The only indication as to the source for the statistics presented is a footnote to one table explaining the difference between one provincial population and that given for the same province the preceding year in the *Anuario* as due to a revision of the vital statistics balance.[4]

The scattered distribution of the population, combined with a widespread lack of cooperation, impaired the value of the census of 1864.[5] The official estimate of the extent of underenumeration in the census of 1892 was 12 percent.[6] The last national census was taken in 1927. The results were published in three volumes, the first devoted to vital statistics, the second to the total populations of provinces and minor areas, and the third primarily to literacy. The preface

---

[1] Thiel, B. A. "Monografía de la población de Costa Rica en el siglo XIX." pp. 3–52 in: *Revista de Costa Rica en el siglo XIX.* Tomo primero. San José, Tipografía nacional, 1902. x, 404 pp.

[2] Provincial populations, reputedly based on censuses, are given for 1844, 1864, 1875, 1883, 1888, and 1892 in the following source: Departamento nacional de estadística. *Resúmenes estadísticos.* I. Sección demográfica, 1883–1889. San José, Tipografía nacional, 1895.

Other sources give provincial statistics from "censuses" in 1824 and 1836. *See*: Dobles-Segreda, Luis. *Indice bibliográfico de Costa Rica, Tomo V, Historia hasta 1900.* San José, Librería y imprenta Lehman, 1933. See especially pp. 576–593, *Censos de Costa Rica.* xiv, 623 pp.

[3] Secretaría de salubridad pública y protección social. Departamento de biodemografía. *Aspectos biodemográficos de la población de Costa Rica.* Informe correspondiente al año 1940 . . . San José, Imprenta nacional, 1942. 183 pp.

[4] Dirección general de estadística. *Población de la República de Costa Rica el 31 de diciembre de 1888.* San José, Tipografía nacional, 1889. 35 pp.

[5] Biolley, Paul. *Costa Rica and her future.* Washington, Judd and Detweiler, 1889. 96 pp. cf., p. 29:
"It is very evident that the figures given, even for the last few years, cannot be considered as exact. The census-taking for the entire Republic presents, indeed, great difficulties. Outside of the plateau central the population is scattered, and the people, still ignorant, do not always lend their assistance in that of which they appreciate neither the purpose nor the utility."

[6] Oficina nacional del censo. *Censo general de la República de Costa Rica . . . 18 de febrero de 1892.* San José, Tipografía y litografía nacional, 1893. ccxvii pp.

to the second volume explained that while the census of 1927 was taken according to the system employed in the United States, innumerable difficulties were encountered in carrying it out, especially poor communication and lack of competent personnel.

No national population census has been taken since 1927, despite the emphasis of the central statistical office on the importance of such a project. Annual estimates have been made on the basis of the net balance of births and in-migration, deaths and out-migration. This technique permits no valid estimates of provincial populations, and in addition yields only rough estimates for the country as a whole.[7] A census of unemployment was taken in 1932, but economy and expediency dictated that it include the unemployed only. Hence it is of little value for population estimates.[8]

Current vital statistics are published in the *Informe* of the *Dirección general de estadística.*[9] More detailed vital statistics were previously published in the *Anuario estadístico,* but there was a considerable hiatus between the period covered by the annual and its date of publication.[10]

COLONIAL AND EARLY NATIONAL CENSUSES AND ESTIMATES

Dobles-Segreda, Luis.
Indice bibliográfico de Costa Rica. Tomo v, Historia hasta 1900. San José, Librería y imprenta Lehman, 1933. 623 pp. D 1451.D63
The section, "Censos de Costa Rica," pp. 576–593, summarizes data on provincial populations from censuses of 1844, 1864, 1875, 1883, 1888, 1892, and 1927. Total populations are given on the basis of censuses in 1824 and 1836.

Thiel, B. A.
"Monografía de la población de Costa Rica en el siglo XIX." Pp. 3–52 in: *Revista de Costa Rica en el siglo XIX.* Tomo primero. San José, Tipografía nacional, 1902. x, 404 pp. F 1543.R45
One table included gives the tribal composition of the Indian population during the period of discovery, 1502–1522 (p. 13) and another table gives the racial composition of the population according to "censuses" of 1522, 1569, 1611, 1700, 1720, 1741, 1751, 1778, and 1801 (p. 8), with detailed information for various of these dates given in the following pages. Data for the 19th century include a summary table giving populations of provinces and sub-provincial areas for censuses of 1824, 1836, 1864, 1875, 1883, 1888, and 1892, with estimates for 1801, 1844, and 1900. Sources of all data are discussed at some length.

NATIONAL POPULATION CENSUSES

CENSUSES OF 1844, 1864, 1875, 1883, 1888

Departamento nacional de estadística.
Resúmenes estadísticos. I. Sección demográfica, 1883–1893. San José, Tipografía nacional, 1895. HA 803. A5 1895
Provincial populations are given according to so-called censuses of 1844, 1864, 1875, 1883, 1888 and 1892. p. 9.

---

[7] Carballo, R. Sergio. "Las actividades estadísticas de Costa Rica." pp. 253–258 in: Inter American Statistical Institute, *op. cit.*

[8] Dirección general de estadística. *Censo de personas sin trabajo, año 1932. . . .* San José, Imprenta nacional, 1933. 35 pp. HD 5734.1932

[9] Dirrección general de estadística. *Informe . . . año 1940.* San José, Imprenta nacional, 1941. 80 pp. Govt. Publ. R. R. HA 802.A7

[10] Dirección general de estadística. *Anuario estadístico, año 1934.* Tomo trigesimoctavo. San José, Imprenta nacional, 1941. 387 pp. HA 802.A2

CENSUS OF 1864

Dirección general de estadística.
Estadística de la población en cuadros demostrativos. San José, 1865. xl, 103 pp.
Not located

Oficina nacional de estadística.
Resúmenes estadísticos, años 1883 á 1910. Comercio, agricultura, industria. San José, Imprenta nacional, 1912. HA 803.A5 1912
See, pp. 118–124: Año 1864. Las profesiones y oficios de los habitantes de la República, dividada en provincias. Este estudio se reproduce "integro del censo general de ese año."

CENSUS OF 1883

Departamento nacional de estadística.
Resúmenes estadísticos . . . I. Sección demográfica 1883–1893. San José, Tipografía nacional, 1895. 275 pp. HA 803.A5 1895
Detailed statistics from the census of 1883 accompany the data for 1892 which are given in this volume.

Dirección general de estadística.
Censo general de la República de Costa Rica, levantado bajo la administración del licenciado Don José J. Rodríguez el 18 de febrero de 1892. San José, Tipografía y litografía nacional, 1893. ccxvii pp. HA 801.A5 1892
Comparative data for 1864 and 1883 are included.

CENSUS OF 1892

Dirección general de estadística.
Censo general de la República de Costa Rica, levantado bajo la administración del licenciado Don José J. Rodríguez el 18 de febrero de 1892. San José, Tipografía y litografía nacional, 1893. ccxvii pp. HA 801.A5 1892

CENSUS OF 1927

Oficina nacional del censo.
Estadística vital, 1906–1925. Natalidad, nupcialidad, mortalidad general, mortalidad infantil. San José, Imprenta Lehmann. 1927. 87 pp.
HA 801.A3 Nos. 1, 2, 3
Población de la República de Costa Rica según el censo general de población, levantado al 11 de mayo de 1927. Por provincias, cantones, y distritos . . . San José, Maria v. de Lines, Librería española, 1927. 20 pp.
Total populations are given for provinces, cantons, districts, and capital cities of the provinces.
Alfabetismo y analfabetismo en Costa Rica según el censo general de población de 11 de mayo de 1927. San José. Imprenta librería y encuadernación Alsina, 1928. 65 pp.
This survey of the literacy of the population nine years of age and over includes classifications of the province, canton, and district populations by sex and age (under 9, 9 or over). Comparative statistics are presented for 1864 and 1892. Historical series are included on school enrollment.

OTHER NATIONAL CENSUSES

AGRICULTURE, 1904–1906

Dirección general de estadística.
Primer censo agrícola general. San José, Tipografía nacional, 1904. 17 pp.
HD 1806.A5 1904
Segundo censo agrícola general. San José, Tipografía nacional, 1905. 17 pp.
HD 1806.A5 1905
Censo general, 1905–1906. Censo agrícola de 1905 San José, Tipografía nacional, 1906. 17 pp. HD 1806.A5 1906

Agriculture, Commerce, and Industry

Dirección general de estadística.
Resúmenes estadísticos, años 1883 a 1910. Comercio, agricultura, industria. San José, Imprenta nacional, 1912. 135 pp. HA 803.A5 1912
Año 1907. Censo comercial. . . . pp. 102–104.
Censos agrícolas, años 1905 y 1910. pp. 107–113.
Año 1907. Censo industrial. pp. 126–133.

COMMERCIAL CENSUS OF 1915

Dirección general de estadística.
Censo comercial, año, 1915. San José, Imprenta nacional, 1917. 210 pp. HF 3251.A4

UNEMPLOYMENT, 1932

Dirección general de estadística.
Censo de personas sin trabajo, año 1932. . . . San José, Imprenta nacional, 1933. 35 pp. HD 5734.A5 1932
Unemployed persons are given by age, sex, duration of unemployment, and customary occupation, by cantons. Persons unemployed or not in the labor force were not recorded.

CURRENT NATIONAL VITAL STATISTICS

(Including Population Estimates)

Dirección general de estadística.
Anuario estadístico, año 1934. Tomo trigesimoctavo. San José, Imprenta nacional, 1941. 387 pp. HA 802.A2
The first major section, pp. 77–109, gives summary statistics on births, deaths, and infant mortality, followed by tables for each province giving for districts the number of births and deaths, by sex; births by legitimacy status; and deaths by broad age groups. Detailed statistics on general and infant mortality follow, pp. 110–268.
Informe. . . . año 1941. San José, Imprenta nacional, 1942. 78 pp. HA 802.A7
Estimated populations and density are given for provinces, 1941. Births are tabulated by source reporting, church or civil, and by sex and legitimacy. Deaths are tabulated by age, presence or absence of medical assistance, and cause. Data on infant mortality are included. Births, deaths, and the vital index are given annually 1900–1941. Persons entering and leaving the country are classified by province.

Secretaría de salubridad pública y protección social. Departamento de biodemografía.
Aspectos biodemográficos de la población de Costa Rica. Informe correspondiente al año 1940. . . . Extracto de la Memoria anual de la Secretaría de salubridad pública y protección social correspondiente a 1940. San José, Imprenta nacional, 1942. 183 pp. RA 191.C8A3
The organization of the vital statistics system of Costa Rica is described. Summary sections follow on the status of the population, causes of death, general comment, and recommendations.

Secretaría de salubridad pública y protección social. Departamento pre-escolar, escolar y educación sanitaria.
Salud. San José, Imprenta nacional, 1937. Irregular. Govt. Publ. R. R.
This periodical is devoted primarily to epidemiology and general public health problems, although it occasionally carries statistics on mortality and morbidity from specific diseases.

CUBA

Historical

Nine censuses were taken in Cuba while that country was a Spanish colony.[1] The first four counts made by the colonial government in 1774, 1792, 1817, and 1827, are thought to be quite inaccurate because of the primitive techniques used in their execution. In addition, a large proportion of the slave population appear to have been concealed by their owners, who feared that the census would be used as a basis for taxation. The general census of 1841 represented an attempt at an actual enumeration rather than merely an estimate based on various records. This census is believed to contain quite accurate statistics on the number of slaves in Cuba, although the census of 1846 again seriously understated their numbers.

The censuses of 1861, 1877, and 1887 were taken in conjunction with Spanish censuses. The statistical system of Spain had been reorganized in 1856. The General Committee of Statistics established at that time attempted to take a census of Spain in 1857. Another census was taken in 1860, and some time in the following year a count was made of the population of Cuba and Puerto Rico. Summaries of these data were published with the Spanish census of 1861 although more detailed statistics were published by the colonial government. Cuba was also included in the censuses of the Spanish Empire for 1877 and 1887. Although these counts were actual enumerations, at least of separate clusters of population, they were subject to serious limitations because of the illiteracy, slavery, and civil disorganization existing in Cuba.

These limitations were overcome to a considerable extent in 1899 when a census was taken under the auspices of the United States during the military occupation. The results of this enumeration, which included agriculture and education, are generally considered the most reliable and comprehensive ever secured for Cuba. Another census was taken by the United States Provisional Government in 1907, but it was less comprehensive. A census taken under the Cuban Republic in 1919 also included fairly detailed population data, but the most recent census, that of 1931, gave only total population by minor civil divisions. In general, the censuses since 1899 have suffered from the tendency to subordinate the entire process of the census to the purpose of securing a basis for electoral allotments.

Vital statistics for the Republic of Cuba reveal serious deficiencies.[2] There are wide fluctuations in the annual numbers of births reported in many areas. No statistical annual is released by the Republic, but a series of mimeographed

[1] The Report of the Census of 1899 included an account and critical examination of all the "censuses" reported as having been taken in Cuba, together with a summary of the results of the actual censuses. This work was reproduced in the Census of 1907. The 1919 census reproduced these earlier statements, as no new sources permitting their correction or completion had been found, and added the results of the census of 1907 to the summary of statistical information from previous censuses. See: Cuba. Dirección general del censo. *Census of the Republic of Cuba, 1919.* Havana, Maza, Arroyo y caso, s. en. c., Printers, [1920?] pp. 263–284. HA 872. 1919. A525

A chronological survey, differing in some details from that above, is given in the following source: Martínez-Fortún, Ortelio, and Martín Díaz, Mariano. "Las estadísticas sanitarias en Cuba." *Eighth American Scientific Congress. Proceedings, Vol. VIII.* Washington, Dept. of Interior 1942. 365 pp.

[2] For the historical development of vital statistics, *see* Martínez-Fortún, Ortelio, and Martín Díaz, Mariano. *Op. cit.*

pamphlets has included data on marriages, births, and deaths through 1935. Information on immigration and emigration is available through 1938.

A census of unemployment was taken in 1939, and a regular population census was to have been taken in 1941. There is no evidence that the latter was taken.

COLONIAL POPULATION CENSUSES

CENSUSES OF 1774, 1792, 1817, 1827, AND 1841.

Cuba. Superintendencia general delegada de real hacienda.

Informe fiscal sobre fomento de la población blanca en la isla de Cuba y emancipación progresiva de la esclava con una breve reseña de las reformas y modificaciones que para conseguirlo convendría establecer en la legislación y constitución coloniales: presentado a la superintendencia general delegada de real hacienda en diciembre de 1844, por el fiscal de la misma. Madrid, Imprenta de J. Martin Alegría, 1845. xviii, 328 pp. F 1763.C95

The following data from the censuses of 1774, 1792, 1817, 1827, and 1841 are included: The proportion of white (free) and colored (slave) persons in the population at the time of each census, and the number of white (free), colored (free) and colored (slave) persons in the censuses of 1827 and 1841.

Cuba. Comisión de estadística.

Cuadro estadístico de la siempre fiel isla de Cuba correspondiente al año de 1827 . . . Habana, 1829, 90 pp. Not located

This appears to have been the principal publication containing the data for the census of 1827.[3]

Resumen del censo de población de la isla de Cuba a fin del año de 1841. Habana, 1842. 68 pp. Not located

See also first citation under Census of 1846.

CENSUS OF 1846

Cuadro estadístico de la siempre fiel isla de Cuba, 1846. . . . Habana, Imprenta del gobierno y capitanía general, 1847. vii, 266, 44 pp. HA 873.1846

A folded table after p. 33 gives the population in 1846 for 12 administrative areas by race, sex, free or slave status, and age. The totals by sex and slave status are compared with similar data for 1841.

CENSUS OF 1861

Spain. Instituto geográfico y estadístico.

Censo de la población de Espana según el recuento verificado en 25 de diciembre de 1860 por la Junta general de estadística. Tomo II, pp. 798–809. Madrid, 1863. HA 1542.1860 (folio)

The summary of returns for Cuba includes population by race, citizenship, sex, marital status, literacy, age in five-year groups, and occupations by race by district.

Cuba. Centro de estadística.

Noticias estadísticas de la isla de Cuba en 1862. . . . Habana, Imprenta del gobierno, capitanía general y real hacienda por S. M., 1864. Pages unnumbered. HA 873.1862

This is the definitive publication of the returns for Cuba of the census taken in 1861 in conjunction with the Spanish census of 1860. Detailed information for individual clusters of population includes race, sex, and free or slave status. An age classification by race and sex is given for the 25 jurisdictional areas.

---

[3] An unofficial source giving more detailed information on the first four censuses is the following: Sagra, Ramón de la, *Historia económico-política y estadística de la isla de Cuba; ó sea de sus progresos en la población, la agricultura, el comercio y las rentas*. Habana, Imprenta de la viudas de Arozoza y Solén, 1831. HC 157.C952

This source gives the population according to the censuses of 1774, 1792, 1817 and 1827 for 18 administrative areas by race, sex, and free or slave status (pp. 3–7). The totals for 1827 agree with those in the *Informe fiscal*. De la Sagra refers to a publication of the results of a census of 1827 which he summarized in his *Anales de ciencias*, Número 32, 1830. This appears to have been the *Cuadro estadístico* cited in this bibliography.

CENSUS OF 1877

Spain. Instituto geográfico y estadístico.

Censo de la población de España, según el empadronamiento hecho el 31 de diciembre de 1877. . . . Tomo I, pp. 679–693. Madrid, Imprenta de la Dirección general del Instituto geográfico y estadístico, 1879–1884. HA 1542.1877

De jure and de facto populations are given for districts by sex for four ethnic groups—Spanish, foreign, Asiatic, and colored.

CENSUS OF 1887

Spain. Instituto geográfico y estadístico.

Censo de la población de España, según el empadronamiento hecho en 31 de diciembre de 1887 . . . Tomo I, pp. 764–771. Madrid, Imprenta de la Dirección general de Instituto geográfico y estadístico, 1891. 2 v. HA 1542.1887 (folio)

De jure and de facto populations, Spanish and foreign, are given by sex for districts, with sub-classifications by sex and race and by literacy and race.

NATIONAL POPULATION CENSUSES

CENSUS OF 1899

U. S. War Department. Cuban Census Office.

Census of Cuba. Bulletin No. I–III. Washington, Govt. Printing Office, 1900. 3 vol. in 1. 24, 15, 17 pp. HA 872 1899.A

Contents: No. 1. Total population by provinces, municipal districts and wards. No. 2. Population by age, sex, race, nativity, conjugal condition, and literacy. No. 3. Citizenship, literacy and education.

Report on the census of Cuba, 1899. Lt. Col. J. P. Sanger . . . Director. Henry Gannett, Walter F. Willcox, Statistical Experts. Washington, Govt. Printing Office, 1900. 786 pp. HA 872 1899.B

Spanish edition has title: *Informe sobre el censo de Cuba, 1899.* Washington, Imprenta del gobierno, 1900. 793 pp. HA 872.1899.B5.

A summary of the returns for earlier censuses are given, pp. 179–181. Other data from whatever census counts were held to be reliable for the characteristic involved are presented in the analytical texts devoted to the distribution of the population by age, sex, race, etc. The history of the non-European elements of the Cuban population is discussed.

Data for the 1899 census, which covered agriculture, education, and sanitation as well as population, are presented in detail.

Cuba. Military Governor, 1899–1902. (Leonard Wood).

Ley electoral municipal adicionada con el censo de población y la ley de perjurio. Habana, Imprenta de la "Gaceta oficial," 1900. 34 pp. English and Spanish text, in alternate sections. JL 1019.A2 1900

Folded tables II–V present condensed information on the age, sex, marital status, birthplace, citizenship, occupation and literacy of the population, including a brief analytical text by J. P. Sanger, Director of the Census.

CENSUS OF 1907

Cuba. Oficina del censo.

Censo de la República de Cuba bajo la administración provisional de los Estados Unidos, 1907. Washington, Oficina del censo de los Estados Unidos, 1908. 707 pp. HA 872 1907.A

Population data presented are similar to those of the 1899 census, covering literacy and occupational data as well as race, sex, age, nativity, marital status, etc. The data from earlier censuses, published in the 1899 census, are repeated, with additions. A full discussion of the evidence for the existence and validity of the earlier censuses is included, and data from the censuses prior to 1899 are given, pp. 169–183.

Cuba: population, history and resources, 1907. Compiled by Victor H. Olmstead, Director, and Henry Gannett, Assistant Director: Census of Cuba taken in the year 1907. Washington, United States Bureau of the Census, 1909. 275 pp. F 1758.C949

". . . a compendium containing data compiled from the census reports of 1899 and 1907, and other reliable sources." Population data are principally as of 1907.

Report of the work performed in the preparation of the municipal registers rendered to Honorable Charles E. Magoon, Provisional Governor of Cuba, by General José de Jesus Monteagudo, Director of the Cuban Census. Havana, 1909. 33 pp. In English and Spanish. JL 1018.A3 1909
Total population, by wards, is given as of the 1907 census.

CENSUS OF 1919

Cuba. Dirección general del censo.
Census of the Republic of Cuba, 1919. Havana, Maza, Arroyo y Caso, s. en. c., Printers, 1920?. xii, 968 pp. HA 872 1919.A525
The Spanish edition has title: *Censo de la República de Cuba. Año de 1919.* Habana, Maza, Arroyo, y Caso, s. en c., impresores, 1920?. xii, 977 pp. HA 872.1919.A52
The text of the 1907 volume covering the existence and validity of early censuses of Cuba is translated. Population data for 1919 are similar to those of the 1907 and 1899 censuses.

Estados que comprenden el número de habitantes y electores de la República, según la enumeración practicada el 15 de septiembre de 1919, publicados en edición extraordinaria de la "Gaceta oficial" de 28 de abril de 1920, por acuerdo de la junta nacional del censo de 9 de abril del propio año, y que se reproducen en este folleto en cumplimiento de dicho acuerdo. Habana, Imprenta y papelería de Rambla, Bouza y cía., 1920. 99 pp. HA 872 1919.A5
Total population and voting population, by minor civil divisions.

Apéndice anual a la memoria del censo decenal verificado en 1919. Habana, Sloane y Fernandez, impresores, 1928. 107 pp. HA 872.A6 1919
Data from previous censuses are summarized, and estimates for the title year presented by province.

CENSUS OF 1931

Cuba. Dirección general del censo.
Censo de 1931. Estados de habitantes y electores. Habana, Carasa y cía., impresores, 1932. 106 pp. HA 872 1931.A5
Total population and electoral population are given for minor civil divisions, with comparable figures for 1919. There are no classifications by race or sex.

OTHER NATIONAL CENSUSES

Cuba. Dirección general de estadística.
Asilos de ancianos, 1936. Havana, 1937. 5 pp. mimeographed. Pan Am. Union
Prepared by the *Sección de migración y trabajo.* This work was done as a preliminary step to the census of unemployment scheduled by the *Secretaría del trabajo.* Classification is by race, sex, nationality, and province.

Cuba. Ministerio de trabajo.
Census of unemployment, 1939. Not located
Unemployed persons in Cuba were classified as workers unemployed throughout the year, workers employed during the sugar season and tobacco harvest only, and small property holders or other independent persons who have been unemployed "for a considerable period of time." For a summary, *see*: *Bulletin of the Pan American Union, 74 : 415.* May, 1940.

CURRENT NATIONAL AND CITY VITAL STATISTICS

(Including Population Estimates)

Comisión nacional de estadística y reformas económicas.
Estadísticas, 1933. Habana, 1935. HA 871.A25
Total population and population density are given by province annually for 1929–1933. Marriages, births, stillbirths and deaths are given for municipalities by race, sex and legitimacy for 1933, with annual totals for Cuba for 1900–1933. Birth rates are charted by race, sex and legitimacy for the individual provinces.

Secretaría de sanidad y beneficencia.

Informe anual sanitario y demográfico de la República de Cuba, correspondiente al año 1934. Sanidad y beneficencia, Boletín oficial, 42 (7–12): 57–69. 1939.
RA 194.C85 A45

Marriages, births, stillbirths and deaths in 1934 are given for districts by race, sex and legitimacy.

Dirección general de estadística.

Autopsiados, 1934. Habana, 1937. 24 pp. Pan Am. Union

Classification is by race, sex, cause of death, occupation, marital status, age, and nationality, for *juzgados municipales.*

Demografía sanitaria, año 1935. Habana, 1937. 10 pp. Pan Am. Union

Statistics on marriages, births, and deaths are given for minor civil divisions by race, sex and legitimacy, with a resume by provinces for 1935 and for the country as a whole for 1900–1935. A short text is included.

Oficina nacional del censo.

Información estadística de la población de la República de Cuba en 30 junio 1935, comparada con la que existía el día del censo (21 de septiembre de 1931). Habana, 1935. Pan Am. Union

A similar report as of December 31, 1935, was issued later. Data include population by race as of the census date and the date in the title, and population for municipalities by race and sex as of the title date.

Dirección general de estadística.

Movimiento de población, año de 1939. Habana, 1939. Pan Am. Union

Population estimates are presented for minor civil divisions.

Ministerio de salubridad y asistencia social.

Salubridad y asistencia social. Boletín oficial 45 (1–4). Jan.-April, 1942.
Govt. Publ. R. R.

Monthly statistics on mortality by cause are published for the city of Havana, Monthly morbidity data are given for the provinces.

DOMINICAN REPUBLIC

Historical

Many estimates of the population of the territory which now constitutes the Dominican Republic were made in the period between 1517 and 1920, the date of the first census.[1] In 1905 the National Congress voted $5,000 for preliminary work on a national census, but no census was taken because of the disturbed conditions in some parts of the country.[2] Hence the census of 1920 was the first national enumeration, although it suffered in both completeness and accuracy from the lack of personnel and the military mobilization which took place in some provinces.[3]

The second census, taken in 1935, covered race, religion, literacy, nationality, labor force, and urban-rural residence. No statistics on the age distribution of the population have been located. The results of this census were published in the issues of the *Anuario estadístico* for the years 1937, 1938, and 1939.

A Central Bureau of Statistics was established in the Dominican Republic in 1905. The task of developing a comprehensive and reliable statistical system was rendered extremely difficult by the previously unorganized condition of official statistics, as well as by severe budgetary limitations. In 1909 an effort was made to clarify the obligations of public and private personnel to supply information to the Bureau, but the political crisis of 1911 resulted in the suspension of all statistical activities for a considerable period of time.[4] The present statistical organization was established in 1934 and 1935.

The Central Statistical Bureau is continuing its endeavors to secure prompt and complete reporting of vital statistics. Detailed tabulations are published in the *Anuario estadístico.*

NATIONAL POPULATION CENSUSES

PROPOSED CENSUS OF 1906

Oficina de estadística nacional.

Informe al ciudadano secretario de estado en los despachos de hacienda y comercio. Santo Domingo, Imp. La Cuna de América, 1906. HA 886.A3

A brief discussion of the census authorized for 1906 is included, p. 23.

---

[1] Oficina de estadística nacional. "República Dominicana, 1937." *Anuario estadístico de la República Dominicana del año 1937.* p. 84. See also: Abad, José Ramon. *La República Dominicana.* Reseña general geográfico-estadística, redactada por José Ramon Abad, por orden del Señor ministro de fomento y obras públicas ciudadana Pedro T. Garrido. Santo Domingo, Imp. de García hermanos, 1888. 400, xxvii pp. 91-93 reproduce estimates covering the period 1789-1888. F 1931.A11.

For the history of the colonization and growth of the population, see: García, José G. *Compendio de la historia de Santo Domingo.* 3. ed., aumentada y corr. Santo Domingo, Impr. de García hermanos, 1893-1906. 4 vol. F 1931.G23.

[2] Tolentino R., Vicente. "Las actividades estadísticas de la República Dominicana." pp. 293-301 in: Inter American Statistical Institute, *op. cit.*

[3] Oficina de estadística nacional. *Ibid.*, p. 85.

[4] Tolentino, R., Vicente. *Op. cit.*

CENSUS OF 1920

Secretaría de interior y policía.

Censo de la República Dominicana . . . Santo Domingo, 1923. xiii, 160 pp. HA 886.A5 1923

Lettered on cover: *Primer censo nacional, 1920.*

CENSUS OF 1935

Dirección general de estadística.

Población de la República Dominicana distribuída por nacionalidades. Cifras del censo nacional de 1935. Ciudad Trujillo, 1937. 19 pp. HB 3062.A5 1935

Population is given by nationality and sex for provinces and communes.

Anuario estadístico de la República Dominicana del año . . . Ciudad Trujillo, 1937–1939. HA 886.A35

Año 1937. Capítulo II. Censo de la República Dominicana, observaciones históricas: pp. 84–111. [A history of population estimates from 1514 to 1920 precedes the presentation of the following data from the 1935 census: Density by province, sex and citizenship status, urban and rural residence, nationality, marital condition, race or color, literacy, religion, labor force, and physical defects.]

Año 1938. Tomo I, pp. 78–151. [The history of population estimates is repeated, as are the census data for 1935 published in the 1937 issue. Additional data presented include the population by sex for communities by urban-rural residence, marital status, nationality, race or color, religion, literacy, and labor force. A graph showing estimated or enumerated total population by five-year periods from 1900 to 1938 is included.]

OTHER NATIONAL CENSUSES

CENSUS OF 1935

Dirección general de estadística.

Anuario estadístico de la República Dominicana del año 1939. Ciudad Trujillo, 1939. HA 886.A35

Tomo I. Censo de habitaciones, 1935. pp. 76–177. Censo agrícola del año 1935, pp. 178–300.

Tomo II. Censo pecuario efectuado el 24 de junio de 1935, pp. 14–174.

INDUSTRIAL CENSUS OF 1936

The Bulletin of the Pan American Union, 72:731–732, December, 1938, cites data from an industrial census taken in 1936. Data included are total number of industrial establishments, total wages paid, and total industrial labor force, by sex and industry. F 1403.B955

CURRENT NATIONAL VITAL STATISTICS

(Including Population Estimates)

Dirección general de estadística.

Anuario estadístico de la República Dominicana, 1940. Ciudad Trujillo, Editoral La Nación, 1941. Tomo I, 854 pp. Tomo II, 883 pp. HA 886.A35

Current vital statistics are usually presented in the first volume of this two-volume series. In the 1940 issue, separate sections are devoted to marriages, divorce, and mortality, pp. 68–414, and fertility, pp. 832–847. Migration statistics are included.

Tomo I, 1941, has been issued.

Crecimiento de la población en la República Dominicana. . . . Ciudad Trujillo, 1937–1942. About 12 pp. HB 3552.A5

This annual publication presents current vital statistics and population estimates for provinces and communes as of Dec. 31. The most recent issue includes estimates as of Dec. 31, 1941.

ECUADOR

Historical

There have been many decrees and plans for censuses throughout the history of Ecuador, but actual census statistics of any type are practically nonexistent. The area was included in the Census of Colombia in 1825.[1] A decree was passed by the National Convention of 1861 providing for a census in 1864 and each succeeding 4 years.[2] At least a partial census appears to have been taken under this authorization in 1864.[3] Executive decrees in 1869, 1870, and 1871 provided for the immediate execution of national censuses, but apparently none were taken.[4] A census was taken in Quito in 1906, but the decrees of the same year providing for censuses of the capitals of provinces remained dormant. Censuses were ordered in both 1929 and 1936, but neither was taken. The director general of statistics stresses the interest of the bureau in the achievement of a census. Since the execution of a national population census was impossible, an attempt was made to carry out a cattle census in 1938, but inadequate funds curtailed even this activity in 1939. There had been some hope that a census might be taken in 1940, but this did not prove to be possible.

Vital statistics are collected and published by the *Dirección general de registro civil,* which also issues yearly releases giving estimated population for provinces, cantons and cities.

NATIONAL POPULATION CENSUSES

CENSUS OF 1825

Republic of Colombia. Secretario de estado.
Memoria. Esposición que el Secretario de estado del despacho . . . al Congreso de 1827. Bogotá, Imprenta de Pedro Cubides, 1827. 36 pp.
J 215.R2 1827

Ecuador was included in this census of Colombia.

CENSUS OF 1861 (PARTIAL)

El Nacional.
Cuadro que manifiesta el censo de la población de la provincia de Esmeraldes." *El Nacional,* Periódico oficial, Época segunda, No. 186. Quito, 5 de abril de 1865.
J 6.E3

"Cuadro que manifiesta el censo de la población de la provincia de Manabí." *Ibid.,* 5 de abril de 1865.

---

[1] A census of 1826, presumably the one of 1825, and various estimates are noted, pp. 163–164 in: Villa Vicencio, Manuel. *Geografía de la República del Ecuador.* New York, Imprenta de Robert Craighead, 1858. 505 pp. J 6.B3

[2] Convención nacional. "Decree sobre el modo de formar el censo jeneral de la población de la República." *El nacional,* Periódico oficial, Número 30, Época segunda, p. 2, Quito, 30 de abril de 1861. J 6.E3.1859

[3] "Cuadro que manifiesta el censo de la población de la provincia de Esmeraldes," and *Ibid.,* "... provincia de Manabí." *El Nacional,* Periódico oficial, Época segunda, No. 186. Quito, 5 de abril de 1865.

[4] Hernández, Augusto A. "Las actividades estadísticas del Ecuador." pp. 307–320 in: Inter American Statistical Institute. *Op. cit.*

CITY CENSUS

QUITO

Dirección general de estadística.
Censo de la población de Quito, 1° de mayo de 1906. Quito, El Comercio, 1906. 16 pp., Tables A–N. HA 1028.Q5A5 1906
Age for minor divisions, under 18, 18–45, 45+; sex; marital status; literacy; place of birth of Ecuadorians born in Quito; nationality of aliens; schools; institutions; convents; monasteries; occupations.

CURRENT NATIONAL POPULATION ESTIMATES

Dirección general de registro civil.
"Población de la República del Ecuador, al 1° de enero de 1941." *Registro oficial 2 (369): 2111–2114.* Nov. 18, 1941. J 6.E3
Estimates for provinces, cantons, and cities.

CURRENT NATIONAL VITAL STATISTICS

Dirección general de estadística.
Boletín general de estadística. Vol. III, No. 6, Nov., 1933. 70 pp. Govt. Publ. P. R.
The section, "Estadística demográfica del año de 1932," pp. 1–52, includes the following: general summary of civil registration statistics, 1932 and 1931, by provinces; births by months, by nationality of fathers, multiple births; marriages for cantons, marriages by civil status, and nationality; deaths by civil status and nationality, for provinces; infant mortality; deaths by cause and age; and hospital and morbidity statistics.
In 1939, this publication was replaced by a monthly bulletin entitled, *Estadística y censos*, which was to summarize all current statistical data for Ecuador. Vol. 1, No. 1, March, 1939, contained a report: "Organización, iniciativas y proceso de trabajos estadísticos y censales en 1938." No copies of this publication have been located.
Informe de la Dirección general de estadística, registro civil y censo al Sr. Ministerio del Ramo. Quito, Tip. L. I. Fernández, 1934. 37 pp. Pan Am. Union
Vital statistics for 1932 and 1933, including causes of death and mortality by age, and migration by nationality.

Ministerio de gobierno.
Informe a la nación del Ministerio de gobierno . . . 1939. Quito, Imprenta del Ministerio de gobierno, 1939. 255 pp. J 225.R4
Summary report of the *Dirección general de registro civil* for 1938, pp. 182–187. The 1942 *Informe* has been issued.

Ministerio de hacienda y crédito público.
Informe del señor Ministerio de hacienda y crédito público al H. Congreso, nacional. Segunda parte, 1941. Quito, Imprenta del Ministerio de hacienda, 1941. 368 pp. HJ 35.A2
"Anexo No. 1, Movimiento demográfico de la República en el año 1940," pp. 239–241, presents summary vital statistics for 1940. "Anexo No. 2, Movimiento demográfico habido en la República del Ecuador en 10 años," 1931–1940, gives births by sex and legitimacy status, deaths by civil status, infant mortality, stillbirths, marriages, divorces, and estimated total populations.

CURRENT CITY VITAL STATISTICS

Guayaquil. Oficina del registro cantonal de la población y estadística.
Síntesis de la estadística demográfica, año 1937. Guayaquil, 1938. 1 p. Estimated population and vital statistics. Pan Am. Union

Guayaquil. Oficina del Registro cantonal de la población y estadística.
Boletín municipal de estadística. Nos. 4–5, 1940. Guayaquil, Imprenta i talleres municipales, 1940. 290, viii pp. HA 1028.G8A3
There are three general sections: "Población," p. 12, "Demografía," pp. 13–32: and "Migración," pp. 33–37. These are followed by detailed tabulations for the individual years 1934–1937 inclusive. No. 6, announced for the end of 1941, will contain the comparable data for 1938, 1939, and 1940.

EL SALVADOR

Historical

The provinces of San Salvador and Sonsonate were included as a part of Guatemala in a poll taken in accordance with the *Real Orden* of Nov. 10, 1776.[1] Some form of count or estimate appears to have been made of the population of the Province of San Salvador in 1807.[2] National censuses were levied in 1878, 1882, 1892, 1901, and 1930.[3] A census was authorized for 1940, but funds were not appropriated for its execution.

The *Dirección general de estadística* was established in 1881. Its early publications were intermittent, but they have appeared regularly since 1912. Vital statistics are published currently in the *Boletín estadístico* and in the *Anuario estadístico.*

PROVINCIAL AND NATIONAL CENSUSES

CENSUS OF 1779

Fonseca, Pedro S.

Demografía salvadoreña. San Salvador, Imprenta Rafael Reyes, 1921. 84 pp.
F 1485 F67

Ch. II, pp. 48–62, "Densidad e incremento de la población," gives provincial populations for the Guatemalan provinces based on the count taken in 1778 in accordance with the Real Orden of 1776. At this time San Salvador and Sonsonate were provinces of Guatemala.

Populations are also given according to censuses of 1878, 1882, 1892, and 1901.

CENSUS (?) OF 1807

El Salvador.

Estado general de la provincia de San Salvador, Reyno de Guatemala, año de 1807. Por Don Antonio Gutiérrez y Ulloa, corregidor intendente de la Provincia. Ediciones de la Biblioteca nacional. San Salvador, Imprenta nacional, 1926. 166 pp.
F 1486 G82

Population by sex, family status, race, and occupations, as of the end of 1807, prepared by Don Antonio Gutiérrez y Ulloa in accordance with the *Real Orden* of Sept. 3, 1807.

ESTIMATES (?) OF 1858–1860

El Salvador.

Estadística general de la República del Salvador, por Lorenzo López. Impresa en la Imprenta del gobierno, en el año de 1858. Ediciones de la Biblioteca nacional . . . San Salvador, Imprenta nacional, 1926. 240 pp. HA 842 1858

Tables are included giving the population by sex, marital status, age (children under 14, by sex 15–50, and total), and occupation, for the Departments of La Paz (1858), Santa Ana (1859), Cuscatlan (1859) and Sonsonate (1860). Similar classifications are also given for some subdivisions within the departments.

---

[1] For a history of the censuses of El Salvador, see: Fonseca, Pedro S. *Demografía salvadoreña.* San Salvador, Rafael Reyes, 1921. 84 pp. F 1485.F67

[2] Gutiérrez y Ulloa, Antonio. *Estado general de la provincia de San Salvador, Reyno de Guatemala, año de 1807.* San Salvador, Imprenta nacional, 1926. 166 pp. F 1486.G82

[3] Brannon, Max P. "Las actividades estadísticas de El Salvador." pp. 468–473 in: Inter American Statistical Institute. *Op. cit. See also: Ibid.,* "Desarrollo histórico de la estadística en El Salvador." pp. 263–278 in: *Eighth American Scientific Congress, Proceedings,* Vol. VIII, Statistics. Washington, Dept. of State, 1942. 365 pp.

CENSUS OF 1878

Dirección general de estadística.
División administrativa y población probable de la República de El Salvador de 1911 . . . San Salvador, Imprenta nacional, 1911. 16 pp. HA 842.A4 1911
The censuses of 1778 and 1878 are cited. The population is given as 161,954 in 1778 and 554,785 in 1878.

CENSUSES OF 1882 AND 1892

No separate publications were located. Summary figures are included in: Fonseca, Pedro S., *op. cit.* The report of the Census of 1901 refers to censuses taken in both 1892 and 1894. See below.

CENSUS OF 1901

Dirección general de estadística.
Boletín de la Dirección general de estadística de la República de El Salvador. San Salvador, Vol. 1, No. 1, Jan. 1, 1902. HA 841.A3
The report of the *Dirección general de estadística* to the *Ministro de fomento* on the general census of March 1, 1901, is included. There is one table, "Resumen del censo de 1901," giving population by sex and race for departments.

CENSUS OF 1930

Dirección general del censo.
Censo de población de 1930. San Salvador, Tip. "La Union" S. S. 1929? 44 p. Pan Am. Union
This pamphlet gives the "Reglamento del censo de población de 1930".

Dirección general de estadística.
Población de la República de El Salvador. Censo de 1° de mayo de 1930. San Salvador, Taller nacional de grabados, 1942. 512 pp. HA 841.A55 1930

CITY CENSUS PUBLICATIONS

Oficina del censo.
Censo de población del municipio de San Salvador levantado el 15 de octubre de 1929. San Salvador, Tipografía La Union, 1930. 63 pp. HA 848. S3A5 1929.
This census of the City of San Salvador was taken in 1929 preparatory to the national census of 1930. The contents are as follows: Part 1, Planning and execution of the census. Part 2, Comments on the census. Part 3, Results (Rural-urban distribution, age and sex composition, nationality, education, etc.); index; results of the classification and analysis. Part 4. Critique of the census.

CURRENT NATIONAL VITAL STATISTICS

(Including Population Estimates)

Dirección general de estadística.
Boletín estadístico No. 22. Dec. 31, 1941. 71 pp. HA 841.A3
"Sección demográfica," pp. 3–28, gives detailed classifications on marriages, births, deaths, and migration for the first half of 1941.
This publication is issued semi-annually.

Boletín estadístico extraordinario, No. 13. Divulgación científica. Dec. 31, 1941. 21 pp. HA 848.S3 1929
Estimated population and urban-rural distribution as of June 30, 1941, p. 6. Summary vital statistics, first half of 1941, by months and departments, p. 7. Migrants registered, first half of 1941, p. 8.
Published irregularly as a supplement to the *Boletín estadístico.*

Anuario estadístico de 1940. Tomo No. 1, Tomo II. San Salvador, Taller nacional de grabados, 1941. Tomo 1, 210, vi pp. Tomo II, 367, v pp. HA 842.A2
The first volume includes the demographic section. Estimated populations are given as of Dec. 31, 1940. Birth, death, and marriage rates are given for 1940, with an appendix giving infant mortality by age and cause, and general mortality by age, sex, and occupation. A considerable portion of the mortality analysis is based, not on rates, but on percentage distributions of all deaths.
Tomo 1, 1941 has been issued.

GUATEMALA

Historical

The earliest reports of the population of the area of Guatemala indicate that it was much more densely settled before the conquest and pacification than at any time throughout the colonial period. The first census of the Kingdom of Guatemala was taken in 1778; data on provincial and village populations were published by Juarros in 1823.[1] Various estimates were made between that period and the first national census of 1880, which revealed a total population of 1,224,602. Other national censuses were taken in 1893, 1921, and 1940. The published reports of the 1940 census indicate the rather extraordinary increase of 63.76 percent in the total population of the country between 1921 and 1940, undoubtedly representing at least in part a more complete enumeration in 1940 than in 1921.

Brief résumés of the reports of the *Dirección general de estadística* are carried in the *Memoria de las labores del ejecutivo en el ramo de hacienda y crédito público,* while the reports of the *Registro civil* are summarized in the *Memoria de las labores del ejecutivo en el ramo de gobernación y justicia.* Detailed classifications of numbers of births and deaths by provinces and minor geographic divisions are carried irregularly in special issues of the *Boletín sanitario de Guatemala.*

A detailed analysis of the population growth and vital statistics of Guatemala was made in connection with the Carnegie Institution's medical survey of the Republic of Guatemala, published in 1938.[2]

NATIONAL POPULATION CENSUSES

CENSUS OF 1880

Dirección general de estadística.

Censo general de la República de Guatemala, levantado en el año de 1880. Guatemala, Estab. tip. de "El Progreso," 1881. 448 pp. HA 813 1880

CENSUS OF 1893

Dirección general de estadística.

Censo general de la República de Guatemala, levantado el 26 de febrero de 1893. Guatemala, Tipografía y encuadernación nacional, 1894. 205 pp. HA 813 1893

CENSUS OF 1921

Dirección general de estadística.

Censo de la población de la República levantado el 28 de agosto de 1921. Guatemala, Talleres Gutenberg, 1924–26. 2 vol. in 3. HA 811.A4

Pt. 1. Población clasificada por municipios, departamentos y zonas, con distinción de población urbana y rural, instrucción, raza, sexo e edades. 1924.

Pt. 2. División política y administrativa, estado civil, nacionalidad, ocupaciones. 1924.

Pt. 2. [pt. 2.] Complemento de la parte II. Estado civil, nacionalidad, ocupaciones, municipio de Guatemala. 1926.

[1] Shattuck, George C., et al. *A medical survey of the Republic of Guatemala.* Washington, Carnegie Institution, 1938. pp. 1–3. RA 814.G 855

See also: Schwartz, Guillermo. *Las actividades estadísticas de Guatemala.* Pp. 327–331 in: Inter American Statistical Institute. *Op. cit.*

[2] Shattuck, George C., et al., *op. cit.* Chs. I and II.

PLANNED CENSUS OF 1930

Dirección de censo.

Census of 1930. Preliminaries. Inspection of the Guatemalan region bounding with Honduras, Republic of Guatemala, Central America. Guatemala, Tipografía nacional, 1932. 116 pp. HA 812.A5 1930a

Censo de 1930. Preliminares. Intereses económicos y comerciales de Guatemala en la región fronteriza con Honduras, Publicaciones de la Comisión de límites. Arbitraje. República de Guatemala, Centro América. Guatemala, Tipografía nacional, 1931. 116 pp. HC 147.G8A5 1930

Reglamento general para la facción del censo de 1930. Guatemala, C. A., Tipografía nacional, 1929. 31 pp. HA 812.A5.1930

CENSUS OF 1940

Dirección general de estadística.

Quinto censo general de población levantado el 7 de abril de 1940. Guatemala, C. A., Tipografía nacional, 1942. 885 pp. Govt. Publ. R. R.

Comparisons are made between the censuses of 1880, 1893, 1921, and 1940 by regions and departments. The following tabulations are included for 1940; total population, by class of population; age by sex; race, religion and language, by sex; education and school age population; marital condition; physical impediments; nationality by sex; and occupations. The majority of the tabulations are given in detail for both departments and municipios.

Reglamento para el censo general de población del año 1940. Guatemala, Tipografía nacional, 1939. 24 pp. Govt. Publ. R. R.

CURRENT CITY CENSUS

Dirección general de estadística.

Análisis del censo urbano de la capital, levantado el 22 de febrero de 1938, considerado en sus cifras globales. Guatemala, C. A., 1939. 31 pp. Bur. of Cen.

An exploratory census of the capital, taken preparatory to the census of 1940.

CURRENT NATIONAL VITAL STATISTICS

(Including Population Estimates)

Dirección general de estadística.

Memoria de los trabajos realizados por la Dirección general de estadística en el año 1934. Guatemala, C. A., Tipografía nacional, 1936. 164 pp. HA 812.A3

There is a brief summary including rates by regions and departments, and tables as follows: "Nosología," año 1934, pp. 2–4. "Divorcios, año 1934," pp. 7–14.

"Nomina de los cuadros estadísticos que se reproducen en este trabajo, proporcionados por la Dirección general de estadística de Guatemala." *Boletín sanitario de Guatemala 10(47): 7–197.* June–Dec., 1939. Govt. Publ. R. R.

These statistics, prepared for the *VIII Congreso panamericano del niño,* celebrated in San José, Costa Rica, Oct., 1939, include population estimates, births, deaths, and natural increase for the period 1935–1938. The majority of the tables concern deaths for provinces, by causes.

Ministerio de gobernación.

Memoria de las labores del ejecutivo en el ramo de gobernación y justicia durante el año administrativo de 1939, presentada a la asamblea legislativa en sus sesiones ordinarias de 1940. Guatemala, C. A., Tipografía nacional, 1940. 392 pp. J 179. R2

Report of Registro civil, pp. 276–279 The 1941 issue is now available.

Ministerio de hacienda y crédito público.

Memoria de las labores del ejecutivo en el ramo de Hacienda y crédito público durante el año administrativo de 1940, presentada a la asamblea legislativa en sus sesiones ordinarias de 1941. Guatemala, C. A., Tipografía nacional, 1941. 793 pp. HJ 18.A4

Report, *Dirección general de estadística,* pp. 665–668.

OTHER CURRENT OFFICIAL POPULATION STATISTICS

Ministerio de hacienda y crédito público.

Estadísticas gráficas, año 1939. Guatemala, Tipografía nacional, 1939. 127 pp. HA 812.A5.1939

HAITI

Historical

Although estimates of the population of Haiti have been made at various times by many individuals and organizations between the sixteenth century and the present, no actual enumeration of the people has ever been made. A so-called census taken in 1918–19 appears to have been an estimate based on informal counts by the communal authorities.[1] The Office of the United States High Commissioner estimated the total population in 1923 at 2,050,000. The National Service of Health of Haiti estimates the present population at 3 million persons.[2] Pending the execution of an actual census, there is no way of estimating even the size of the total population within a wide margin of error.

A statistical bulletin of the Republic published by Antoine Laforest in 1913 contains figures for 24 communes which he states are the preliminary results of a census which had already been in progress for over a year.[3] No final results of this count have been located.

Vital statistics, published by the National Health Service, are based on a civil registration system. The extent of under-enumeration is obviously great.[4] For instance, the recorded death rate per 1,000 estimated population was 3.4 in 1940 and 3.1 in 1941. The National Health Service was reorganized in 1940, with the goal of improving the civil registration system. However, the public health budget of Haiti was reduced for 1940–1941, with the result that public health services had to be contracted.[5]

Because the great bulk of the Haitian population belongs to the Catholic Church, parochial estimates of population and church records of baptisms, marriages and interments constitute an important source of information on the Haitian population. Fortunately, the boundaries of the dioceses of the Catholic Church in Haiti correspond fairly well to the boundaries of the administrative departments of the Republic.

NATIONAL POPULATION "CENSUSES" AND ESTIMATES

Haiti.

Bulletin des statistiques de la République d'Haiti, publié par Antoine Laforest. Octobre 1912–Septembre 1913. Port-au-Prince, Imprimerie de l'Abeille, 1913. 316 pp. HA 881.A3

This volume, published under the auspices of the *Secrétaire d'état de l'intérieur*, contains a note on population, pp. 8–9. According to this note, no count or census had ever been taken. The Department of the Interior had been attempting

---

[1] League of Nations, Economic intelligence service. *Statistical Yearbook of the League of Nations, 1939–40.* Geneva, 1940. 285 pp. Table 2, p. 15.

[2] Thebaud, Jules. La sante publique en Haiti. *Boletín de la Oficina sanitaria Panamericana 22 (1): 20–22* Enero, 1943.

[3] *Bulletin des statistiques de la République d'Haiti*, publie par Antoine Laforest. Octobre 1912–Septembre 1913. Port-au-Prince, Imprimerie de l'Abeille, 1913. 316 pp.

[4] Service national d'hygiène et d'assistance publique. *Rapport annuel du directeur général, 1939–40.* Port-au-Prince, 1941. Pan American Union Library

The director states, p. 139: "En attendant, nous continuons, comme par le passé, à présenter des statistiques de vitalité et de morbidité qui, en ce qui concerne les informations fournies par les officiers de l'État-Civil, restent bien au-dessous de la vérité."

[5] Thebaud, Jules, *op. cit.*

to take a census for over a year, but only 24 communes were covered, not including Port-au-Prince. The population for these 24 communes is reported, generally by urban-rural residence.

Estimated Catholic populations by diocese are quoted for 1912, and the total Protestant population estimated. The total thus yielded is compared with a private estimate of the total population.

Vital statistics for Port-au-Prince for the title year are given for three areas of the city by month, births being reported by sex and legitimacy status, and deaths by sex and adult-child classification.

United Kingdom. Department of Overseas Trade.

Report on economic conditions in the Dominican Republic and in the Republic of Haiti. London, H. M. Stationery Office. 19– to date. HC 157 S2G7

These reports, usually written by the consuls at Port-au-Prince, include population estimates and brief discussions. The issue for 1924–1925 estimates the population at 2½ million, but states that "a recent publication by the clergy gives the figure as 2,628,000." This same issue reproduces estimates of the population of eleven towns, crediting them to the Gendarmerie d'Haiti. The issue for 1936, the first report issued after the evacuation of the American forces, describes the racial and social composition of the population. Total population is estimated at "in the neighbourhood of three million."

U. S. Congress. Senate. Select Committee on Haiti and Santo Domingo.

Hearing before a Select Committee . . . Sixty-seventh Congress, First Session, Pursuant to S. Res. 112, authorizing a special committee to inquire into the occupation and administration of the territories of the Republic of Haiti and the Dominican Republic. Part I, Aug. 5, 1921. Part 2, Oct. 4 to Nov. 16, 1921. Part 3. Nov. 29, 1921. Washington, Govt. Printing Office, 1922. 1842 pp. F 1926.U54

The report by Carl Kelsey, *The American intervention in Haiti and the Dominican Republic*, pp. 1279–1315, states that no census has ever been taken. Population is estimated at 2,000,000, as contrasted to a population of 550,000 in 1880. This source also gives summary descriptions of origin, distribution, towns, health, and economy.

U. S. High Commissioner to Haiti.

Data on the physical features and political, financial and economic conditions of the Republic of Haiti. American High Commissioner, Republic of Haiti, April 1923. 8 pp. F 1926.U573

Total population, population density, interior population, and population of the leading eight cities are estimated.

Later issues of the High Commissioner's report include information on health conditions and activities, but none on population *per se.*

Province ecclésiastique d'Haiti.

Le bulletin de la quinzaine, paraissant tous les quinze jours. 15 ème année, Nos. 19–20. Dimanche, 22 Janvier, 1939. Statistique générale due clergé et des congrégations religieuses de la Province ecclésiastique d'Haiti au 22 Janvier 1939. Private Library

The Catholic population of the five dioceses of Haiti is given, p. 240. These diocesan populations can be taken as approximations of departmental populations because more than 95 per cent of the inhabitants are members of the Catholic Church, and the boundaries of the dioceses correspond fairly well to the departmental boundaries. Population of individual parishes is given, pp. 243–246.

CURRENT NATIONAL VITAL STATISTICS

(Including Population Estimates)

Haiti. Service national d'hygiène et d'assistance publique.

Rapport annuel du directeur général, 1940–1941. Port-au-Prince, 1942. 197 pp. RA 194.H2A3

This report has been published intermittently since 1928. It includes estimated total populations, vital statistics compiled from the Civil Register, and morbidity and mortality statistics based on hospital records.

The statistics included in these annual reports reveal the inadequacies of official statistics. If the "census" population of 1,631,000 in 1918–1919 and the official estimate of 3,000,000 in 1940 are accepted, the geometric rate of increase would be almost three per cent. In 1940, there were only 45.6 thousand registered births and 14.8 thousand registered deaths, whereas a population of 2.5 million with a birth rate of 40 and a death rate of 20 would yield 100 thousand births and 50 thousand deaths. [Statistics given by Giorgio Mortara in: Revista brasileira de estatística 2 (7). July–Sept., 1941.]

LOCAL VITAL STATISTICS

Archidiocèse de Port-au-Prince.

Annuaire de l'Archidiocèse de Port-au-Prince, 1939. Port-au-Prince.
Prinak Library.

Annual vital statistics for the Catholic population include the number of baptisms, marriages and interments, by parish.

OTHER CURRENT NATIONAL POPULATION STATISTICS

Duvivier, Ulrick.

Bibliographie générale et méthodique d'Haiti. Tome I, II. Port-au-Prince, Imprimerie de l'état, 1941. Tome I, 318 pp. Tome II, 411 pp. Z 1531 D.88

Sections referring to population materials are as follows: Ethnologie-Ethnographie, Tome I, pp. 74–80. Sciences medicales, Tome II, pp. 169–206.

Vincent, Stenio. President d'Haiti.

Efforts et résultats. Port-au-Prince. Imprimerie de l'état, 1938. F 1926.V76

A discussion of the problems with which the new government of Haiti is attempting to deal and of the measures taken to solve them. There are sections dealing with land ownership, education, and health problems.

HONDURAS

Historical

The Republic of Honduras has a long history of both population estimates and censuses. Soon after Señor Obispo Fray Fernando de Cadiñanos arrived in 1788 as head of the Diocese of Comayagua, he started a series of systematic visits to all parts of the diocese. His counts of settlements and souls were transmitted to the Real and Supreme Council of Indias; they constitute the census of 1791. The Governor of the Province, Don Ramón de Anguiano, was responsible for another census in 1801, which included lists of settlements, and the number of Spanish and Spanish-speaking families, unmarried men and Indian taxpayers in each. The majority of the Indians were omitted.[1]

Estimates are available for 1826 and 1850. The first approximation to a census in the modern meaning of the term was that taken in 1881 under the direction of Francisco Cruz as chief of the recently organized *Dirección general de estadística*. However, this census was not levied systematically department by department.[2] Official sources refer to a count of 1887, but only detailed estimates of the population as of December 31, 1888, were located. A census of 1895 is also mentioned.[3]

A census was attempted in 1901, but its coverage appeared so incomplete that the statistical office refused to accept it without adjustments. Inspection revealed that in cities where the census was thought to be good the number of men in military service and personal service on road work constituted a fairly regular proportion of the total population. This proportionate relationship was then used to estimate the total populations of the various areas.[4] Another census, taken in 1905, yielded a smaller population than the corrected figures for 1901. The director of the 1905 census consequently based his comparisons on the actual returns in both 1905 and 1901.[5]

National censuses with a somewhat greater degree of validity were taken in 1910, 1916, and 1926. Quinquennial censuses have been taken since 1930.

NATIONAL POPULATION CENSUSES

EIGHTEENTH AND NINETEENTH CENTURY CENSUSES

Cruz-Zambrano, Miguel A.
"Las actividades estadísticas de Honduras." pp. 343–349 in: InterAmerican Statistical Institute, *op. cit.* Washington, 1941. 842 pp. HA 175.S75

---

[1] Dirección general de estadística. *Breve noticia del empadronamiento general de casas y habitantes de la República de Honduras practicado el 18 de diciembre de 1910.* Tegucigalpa, Tipografía nacional, 1911. (HA821.A5.1910

See also: Cruz Zambrano, Miguel A. "Las actividades estadísticas de Honduras." pp. 343–349. In: Inter American Statistical Institute. *Op. cit.* HA 175.S75

[2] Dirección general de estadística. *Cuaderno número I que contiene el movimiento de población correspondiente al año de 1888.* Tegucigalpa, Tipografía del gobierno, 1890. 31 pp. HA 821.A4

[3] *Ibid*, p. 31.

[4] Dirección general de estadística. *La población de Honduras en 1901.* Tegucigalpa, Tipografía nacional, 1902. 71 pp. HA 821.A5.1902

[5] Dirección general de estadística. *La población de Honduras en 1905.* Tegucigalpa, Tipografía nacional, 1906. 13 pp. HA 821.A5.1906

Dirección general de estadística.

Breve noticia del empadronamiento general de casas y habitantes de la República de Honduras practicado el 18 de diciembre de 1910. Tegucigalpa, Tipografía nacional, 1911. 28 pp. HA 821.A5 1910

Pages 6–7 summarize the development of the population of Honduras from 1791 through 1910, according to the registrations, estimates, censuses and counts between the two years. Numbers of inhabitants are given for 1791, 1801, 1826, 1850, 1881, 1887, 1895, 1901, 1905, and 1910.

Cuaderno número I que contiene el movimiento de población correspondiente al año de 1888. Tegucigalpa. Tipografía del gobierno, 1890. 31 pp. HA 821.A4

Data on and a brief critique of the census of 1881.

CENSUS OF 1901

Dirección general de estadística.

La población de Honduras en 1901. Tegucigalpa, Tipografía nacional, 1902. 71 pp. HA 821.A5 1902

CENSUS OF 1905

Dirección general de estadística.

La población de Honduras en 1905. Tegucigalpa, Tipografía nacional, 1906[?]. 13 pp. HA 821.A5 1906

CENSUS OF 1910

Dirección general de estadística.

Breve noticia del empadronamiento general de casas y habitantes de la República de Honduras, practicado el 18 de diciembre de 1910. Tegucigalpa, Tipografía nacional, 1911. 28 pp. HA 821.A5 1910

A summary of the development of the population from 1791 through 1910 is included. Registrations, estimates, counts or censuses are reported as having been taken in 1791, 1801, 1826, 1850, 1881, 1887, 1895, 1901, 1905, and 1910.

CENSUS OF 1916

Dirección general de estadística.

Censo general de población. [Dec. 17, 1916] pp. 3, 9–93 in: Informe del señor director general de estadística nacional al señor Ministro de gobernación y justicia, 1916. Tegucigalpa, Tipografía nacional, 1919. 763 pp. HA 821.A35 1916

CENSUS OF 1926

Dirección general de estadística.

Censo general de población. 1927. Tegucigalpa, Tipografía nacional, 1928. 126 pp. HA 821.A5 1928

The title page has the date 1927, but the table headings report the census as having been taken on Dec. 26, 1926.

CENSUS OF 1930

Dirección general de estadística.

Resumen del censo general de población levantado el 29 de junio de 1930. Tegucigalpa, Tipografía nacional, 1932. 202 pp. HA 821.A5 1930

CENSUS OF 1935

Dirección general de estadística.

Resumen del censo general de población levantado el 30 de junio de 1935. Tegucigalpa, Talleres tipográficos nacionales, 1936. 205 pp. HA 821.A5 1935

CENSUS OF 1940

Dirección general de estadística.

Resumen del censo general de población levantado el 30 de junio de 1940. Tegucigalpa, Talleres tipográficos nacionales, 1942. 214 pp. HA 821.A5 1940

Statistics are given for major and minor civil divisions, age and sex, urban and rural distributions, civil status, race, nationality, occupation, religion, education, and literacy.

An analysis of trends, 1881, 1887, 1910, 1916, 1926, 1935, and 1940 is included.

CURRENT NATIONAL VITAL STATISTICS

Dirección general de sanidad.

Boletín sanitario. Revista trimestral. Año 1, Jan., 1926. Last available, Año VIII, Nos. 31 y 32, June 1, 1941. RA 191.H6A25

Recent issues have carried a section entitled: "Datos demográficos de la República de Honduras durante los años comprendidos de . . . a . . ." Data include births by sex and legitimacy status, deaths, infant mortality, stillbirths and marriages.

Secretaría de gobernación, justicia, sanidad y beneficencia.

Informe de los actos realizados por el poder ejecutivo en los ramos de gobernación, justicia, sanidad, y beneficencia. Presentado al Congreso nacional por el secretario de estado Ing. Abraham Williams, año fiscal de 1940 a 1941. Tegucigalpa, 1941. 281 pp. J 181.R2

Vital statistics, primarily for 1940 and 1941, are given in the following section: "Estadística nacional, Informe de la labor realizado en la Dirección general de estadística durante el año económico de 1940 a 1941." Births are classified by color and legitimacy status; marriages by marital status, nationality, and age; and deaths by age, marital status, religion, occupation and cause. Tabulations are by departments. Migration statistics are included.

There is also a section on population, "Resumen del censo general de población levantado el 30 de junio de 1940." pp. 161–172.

MEXICO

Historical

Many attempts to estimate or count the population of Mexico were made throughout the first three centuries of the colonial period, although only fragmentary records exist for most of them.[1] The first official survey was made in 1579–82, when Philip II of Spain ordered an inventory of the physical, natural, social, and economic resources of the territories under his rule. The original manuscripts of this census are in the Mirabeau B. Lamar Library of the University of Texas.

Three surveys are known to have been prepared in the latter part of the sixteenth century, while eight were made in the seventeenth century, in 1607–10, 1614, 1625, 1652, 1662, 1664, and 1655. None of these were published. A general inventory of the population and economy of the country was undertaken in 1742. However, the first comprehensive attempt at an actual enumeration of the population came in 1791. Returns were incomplete both as to area and coverage, and the results were never published in any detail.[2] Many estimates were made thoughout the nineteenth century, but no attempt was made to take another census until 1895, when the first of the 6 national censuses was taken. The results of this census were published in considerable detail in some 30 volumes, although inadequate preparation, lack of cooperation on the part of large parts of the native Indian population, and similar factors impaired the accuracy of both the census of 1895 and that of 1900. The census of 1921 was taken according to the European system, with the heads of the individual families filling out the schedules. The normal difficulties of securing accurate census statistics during this period were further complicated by the aftermath of the past decade of violent political upheaval.

The *Dirección general de estadística*, established in 1923, was in charge of the planning and execution of the 1930 census. Careful plans were made, and an experimental census taken in the State of Morelos in 1929. The problems which emerged were then discussed at the Second Statistical Conference in Mexico City in 1930.[3] The result was that the returns of this census were far more complete and accurate than those of any preceding census in Mexican history.

The development of census techniques and materials in the last decade has not been limited to the population field. The census law of May 23, 1938, provided for a decennial population census, an agricultural-livestock census in the years

---

[1] This summary statement relies heavily on the section on Mexico, pp. 773–776, in: *Economic Literature of Latin America*. Detailed bibliographies of colonial and national population and vital statistics publications are included in the *Bibliografía mexicana de estadística*, Tomo 1, published by the Dirección general de estadística, 1941. See also: Bojórguez, Juan de Dios. "Las actividades estadísticas de México." pp. 359–376 in: Inter American Statistical Institute. *Op. cit.* A study based on the original census records and computation sheets of the Census of 1790 has just been published. See: Cook, S. F. "The population of Mexico in 1793." *Human Biology 14 (4): 499–515.* Dec., 1942. GN I.H8

[2] Forty volumes of the returns are preserved in the *Archivo de la nación* in Mexico City. The records of local censuses taken in 1796, 1800, 1801, 1811, 1812, and 1813 are also housed in the *Archivo de la nación*.

[3] Departamento de la estadística nacional. *Segunda reunión nacional de estadística*. 20 de diciembre de 1929 a 8 de enero de 1930. México, D. F., 1930. 262 pp. HA 12.5.R4

537846—43——5

ending in zero, a decennial census of buildings in the years ending in nine, and industrial, commercial, transport, and land censuses every 5 years.[4] The last population census, that of March 1940, was part of a general census which included the Sixth Census of Population, the Second Census of Agriculture and Livestock; the Second Public Land Census; the Second Building Census; the Third Industrial Census; the First Business Census; and the First Transportation Census.[5] Preliminary returns of these various censuses were published currently in the Revista de estadística, the monthly statistical bulletin of the *Dirección general de estadística.*

Vital statistics for Mexico are available for the 25 years between 1886 and 1910, but are lacking for the period between 1911 and 1921. Since 1922 the Mexican Government has devoted increasing attention to the problems involved in securing accurate and comprehensive statistics. Despite the general improvement during the last two decades, under-registration of births still constitutes a problem and the validity of the statistics on causes of death for rural areas is questionable. Current vital statistics are published in the *Anuario estadístico* and in the monthly *Revista de estadística.* The former publication contains definitive figures, with detailed classifications for the individual States, while the latter usually contains national data only. It includes some preliminary data for a later period than that covered by the *Anuario.*[6]

NATIONAL POPULATION CENSUSES

COLONIAL AND EARLY NATIONAL ESTIMATES AND CENSUSES [7]

Bojórquez, Juan de Dios.
"Las actividades estadísticas de México." pp. 359–376 in: Inter American Statistical Institute, *op. cit.* Washington, 1941. 842 pp. HA 175 S75

Dirección general de estadística.
Bibliografía mexicana de estadística. Tomo 1. Generalidades, teoría y aplicaciones metodológicas, demografía, estadística social, estadística económica, estadística administrativa, geografía. México, D. F., 1942. 696 pp. Z 7554.M6M6

The section on demography, pp. 143–181, includes 503 titles on status, vital statistics, migration, and population policy.

---

[4] Inter American Statistical Institute. *op. cit.*, p. 352.

[5] Agricultural and livestock censuses were taken in 1930 and 1940; volumes for the individual States for 1930 are still in process of publication, and only preliminary returns are available for 1940. Building censuses, which included housing data, were taken in 1929 and 1939, and have been published in full. Ejidal censuses, covering the twelve months preceding the census data, were carried out in 1935 and 1940 by the *Dirección general de estadística* in cooperation with the *Departmento agrario,* the *Secretaría de agricultura y fomento,* and the *Banco nacional de crédito.* The industrial censuses of 1930, 1935, and 1940 were carried out by a combination of personal interview and correspondence. The censuses of 1935 and 1940 covered only those concerns which had produced goods valued at $10,000 or more during the previous year.

[6] Statistics on immigation and emigration have been gathered since 1908, but the classifications from year to year are not comparable. Detailed historical statistics are published in the *Anuario estadístico.* In 1939 the government began the collection of data on the permanency of tourists entering and leaving the country. A special series on this subject is included in the *Revista de estadística.*

[7] Estimates and counts are available for various local areas. Data indicating the population growth of the Federal District from 1524 to 1940 are presented in the following publication: Departamento del Distrito Federal. Oficina de estadística y estudios económicos. *Memoria del 1° de septiembre de 1940 al 31 de agosto de 1941.* México, D. F., Talleres gráficos de la Penitenciaría del Distrito Federal, 1941. Govt. Publ. R. R.

Anuario estadístico de los Estados Unidos Mexicanos, 1940. México, D. F., 1942. 806 pp. HA 762.A3

Section II, *Población del país*, contains historical estimates and data from the various censuses. The discussion of sources of early population estimates and censuses includes a tabular compilation of total population from 1521 to 1940.

Harvard University. Bureau for Economic Research in Latin America.

The economic literature of Latin America, a tentative bibliography. Vol. II. Cambridge, Harvard University Press, 1936. HA 175 S5

Population sections republished in: Inter American Statistical Institute. *op. cit.*, pp. 741–805. *See especially*, pp. 773–777.

CENSUS OF 1895

Dirección general de estadística.

Censo general de la República Mexicana verificado el 20 de octubre de 1895. Mexico, D. F., Oficina tip. de la Secretaría de fomento, 1897–1899. 30 vol. HA 761 1895.A2

There is a separate volume for each of the following states: Aguascalientes, Baja California, Campeche, Chiapas, Chihuahua, Coahuila, Colima, Distrito Federal, Durango, Guanajuato, Guerrero, Hidalgo, Jalisco, México, Michoacán, Morelos, Nuevo León, Oaxaca, Puebla, Querétaro, San Luis Potosí, Sinaloa, Sonora, Tabasco, Tamaulipas, Tepic, Tlaxcala, Veracruz, Yucatán, Zacatecas.

Censo general de la República Mexicana verificado el 20 de octubre de 1895. Resumen. México, D. F., Oficina tip. de la Secretaría de fomento, 1899. 502 pp. HA 761 1895.A3

CENSUS OF 1900

Dirección general de estadística.

Censo general de la República Mexicana verificado el 28 de octubre de 1900. . . México, D. F., Oficina tip. de la Secretaría de fomento, 1901–1904. 32 vol. HA 761 1900.A2

There are three volumes for Oaxaca, and one for each of the other states.

Censo general de la República Mexicana practicado en 1900. Extranjeros residentes. México, D. F., Oficina tip. de la Secretaría de fomento, 1903. 225 pp. HA 761 1900.A5

Resumen general del censo de la República Mexicana, verificado el 28 de octubre de 1900. México, D. F., Imprenta y fototipía de la Secretaría de fomento, 1905. 79 pp. HA 761 1900.A4

CENSUS OF 1910

Dirección general de estadística.

Tercer censo de población de los Estados Unidos Mexicanos verificado el 27 de octubre de 1910. México, D. F., Oficina impresora de la Secretaría de hacienda, Departamento de fomento . . . , 1918–1920. 3 vol. HA 761 1910.A5

CENSUS OF 1921 [8]

Dirección general de estadística.

Censo general de habitantes. 30 de noviembre de 1921. México, D. F., Talleres gráficos de la nación, 1925–1928. 31 vol. HA 761 1921.A3

The data for Baja California, Territorio Norte and Baja California, Territorio Sur are combined in one volume. There is one volume for each of the remaining thirty states.

Resumen del censo general de habitantes de 30 de noviembre de 1921. México, D. F., Talleres gráficos de la nación, 1928. 203 pp. HA 761 1921.A5

CENSUS OF 1930

Departamento de la estadística nacional.

Memoria de los censos generales de población, agrícola, ganadero e industrial de 1930. . . . México, D. F., Talleres gráficos de la nación, 1932. 212 pp. HA 37.M7 1930

This volume on the organization and execution of the 1930 census is prefaced by a discussion of earlier censuses and pre-census estimates. Figures for the total population are supplied for 1793–1921. The experimental census of the state of Morelos is discussed, and some of the results are given. No results are given for the general census of 1930.

Dirección general de estadística.

Quinto censo de población, 15 de mayo de 1930. México, D. F., Talleres gráficos de la nación, 1932–1936. 32 vol. (Baja California, Distrito Norte, Campeche and Coahuila have imprint: Editorial Cultura; Aguascalientes, Baja California, Distrito Sur, Nayarit and Puebla, Cía. imp. papelería S. A.) Govt. Publ. R. R.

A separate volume was issued for each State as follows: Aguascalientes. 1933. 63 pp. Baja California, Distrito Norte, 1933, 59 pp. Baja California, Distrito Sur. 1933. 71 pp. Campeche. 1934. 65 pp. Coahuila. 1933. 99 pp. Colima. 1934. 59 pp. Chiapas. 1935. 363 pp. Chihuahua. 1935. 235 pp. Distrito Federal. 1932. 83 pp. Durango. 1936. 185 pp. Guanajuato. 1935. 233 pp. Guerrero. 1934. 195 pp. Hidalgo. 1936. 209 pp. Jalisco. 1936. 507 pp. México. 1933. 135 pp. Michoacán. 1935. 329 pp. Morelos. 1935. 69 pp. Nayarit. 1933. 103 pp. Nuevo León. 1934. 201 pp. Oaxaca. 1936. 549 pp. Puebla. S. A., 1933. 149 pp. Querétaro. 1935. 81 pp. Territorio de Guintana Roo. 1935. 49 pp. San Luis Potosí. 1935. 185 pp. Sinaloa. 1935. 165 pp. Sonora. 1934. 195 pp. Tabasco. 1935. 60 pp. Tamaulipas. 1935. 197 pp. Tlaxacala. 1935. 81 pp. Veracruz. 1936. 453 pp. Yucatán. 1934. 185 pp. Zacatecas. 1935. 38 pp.

---

[8] The following publications of the *Dirección general de estadística* containing the instructions for the census of 1921 were not located:

Censo de 1921. Disposiciones dictadas por la Dirección general de estadística para organizar los trabajos preparatorios del censo general de habitantes que deberá verificarse el 30 de noviembre de 1921. México, D. F., 30 pp.

Censo de 1921. Cuarto censo general de la población. Instrucciones sobre la ejecución de los trabajos censales. México, D. F., 1921. 31 pp.

Censo de 1921. Primera concentración de las cédulas para habitantes. Instrucciones dirigidas a las secciones de estadística, con un apéndice relativo a la nomenclatura de ocupaciones. México, D. F., 1922. 90 pp.

[9] The following additional publications were issued by the *Departamento de la estadística nacional, Dirección de los censos:*

Instrucciones para empadronadores, jefes de manzana, de sección, de cuartel y agencias censales. México, D. F., 1929. 16 pp. Not located.

Instrucciones generales para la ejecución de los censos de población y agrícola-ganadero, 15 de mayo de 1930. 40 pp. Not located.

Introducción a la memoria de los censos de 1930, por el Ing. Juan de D. Bojórquez . . . México, D. F., 1930. 46 pp. Pan American Union.

The *Secretaría de educación pública* also issued a volume of instructions: *Instrucciones a los C. C. directores de educación federal, profesores, inspectores y maestros primarios y rurales, para que sustenten pláticas ilustrativas sobre los próximos censos de población, agrícola-ganadero, e industrial de 15 de mayo de 1930.* México, D. F., 1930. Pan American Union.

A preliminary experimental census was taken in 1929 in the state of Morelos. No separate publication was located. The *Departamento de la estadística* issued the following pamphlet: *Instrucciones para la ejecución de los censos en el estado de Morelos, 25 de julio de 1929.* México, D. F., 1929. 9 pp.

There is a uniform plan of presentation for each volume. Population is given by sex for county and township units. Distribution by size of place is given for census years from 1900 to 1930, and age distribution by quinquennial age groups by sex for 1921 and 1930. Marital status is given by sex and age for 1930, and literacy for persons ten years old and over is given by sex for counties. The total is classified by occupation and sex for 1921 and 1930 and by nationality and sex for census years from 1900. Classification by place of birth is for 1921 and 1930, by sex. Language spoken is recorded for all persons five years old and over. The population is further classified by size of family and the presence of physical or mental defects. Each volume contains a copy of the census schedule. [10]

Quinto censo de población, 15 de mayo de 1930. Resumen general. México, D. F., Talleres gráficos de la nación, 1934. xxxi, 269 pp. Govt. Publ. R. R.
Classifications used in this volume are in general the same as those in the individual state volumes, except that few characteristics are given for units smaller than the state. Labor force status is given by sex for each state for census years from 1900 to 1930.

Quinto censo de población, 15 de mayo de 1930. Población municipal. . . . México, D. F., Talleres gráficos de la nación, 1934. 38 pp. HA 761 1930.A5
State summaries of municipal populations, by sex only.

CENSUS OF 1940 [11]

Laws, statutes, etc.
Decreto que declara de interés nacional la organización y levantamiento de los censos 1939–1940. *Diario oficial 113 (22) 1–3.* March 25, 1939. J 4.A3
This decree established the legal and administrative framework for the various censuses taken in 1939–1940.

Dirección general de estadística.
Compendio estadístico. México, D. F., Talleres gráficos de la nación, 1941. 117 pp. HA 37.M7 1941
Preliminary total populations for states, 1940.

Anuario estadístico de los Estados Unidos Mexicanos, 1940. México, D. F., 1942. 806 pp. HA 762.A3
The second section, Población del país, pp. 20–84, includes preliminary total populations for states, counties, and cities of 20,000 or more inhabitants.

"Población extranjera en el Distrito Federal." Censo de 1940. Departamento del Distrito Federal, *Boletín de estadística*, p. 25. Dec. 1941. Govt. Publ. R. R.
The foreign population of the Federal Districts and its subdivisions is classified by sex.

Sexto censo general de población de los Estados Unidos mexicanos, 6 de marzo de 1940. Población municipal. México, D. F., Talleres gráficos de la nación, 1943. 56 pp. Govt. Publ. R. R.

OTHER NATIONAL CENSUSES [12]

AGRICULTURE, 1930

Dirección general de estadística.
Primer censo agrícola-ganadero, 1930. México, D. F., Talleres gráficos de la nación. 1936–19—. Govt. Publ. R. R.
Vol. I. Resumen general.
Vol. II. Parts for Aguascalientes, Campeche, Coahuila, Distrito Federal, Guanajuato, Hidalgo, Jalisco, México, Michoacán, Morelos, Puebla, Querétaro, San Luis Potosí and Tlaxcala.

---

[10] Final results were published also in the *Diario oficial. See:* "Censo de población de los Estados Unidos Mexicanos el 15 de mayo de 1930." *Diario oficial 82(6): 98–99.* Jan. 8, 1934. J4. A3
(Final results for state populations by sex)

[11] The following publications of the *Dirección general de estadística* on the occupational parts of the 1940 census were not located. *Nomenclatura nacional de ocupaciones: 1940.* México, 1941. 1062 pp. *Utilidad de los datos estadísticas sobre ocupación, censos nacionales 1939–1940.* México, D. F., 1941.

[12] Data from various of these special censuses have been included in the following publications of the *Dirección general de estadística:*
Anuario estadístico de los Estados Unidos Mexicanos 1940. México, D. F., 1942. 806 pp. HA 762.A3
Compendio estadístico. México, D. F., 1941. 117 pp. HA 37.M7. 1941

AGRICULTURE, 1940

Dirección general de estadística.
Censo agrícola-ganadero de 1940. "Predios de 1 a 5 y menores de 1 hectárea." *Revista de estadística 4 (3): 195–196.* June, 1941. Govt. Publ. R. R.
"Predios agrícolas de más de 5 hectáreas, . . . censo de 1930." *Ibid.* 5 (1): 3. Jan., 1942.
"Principales características de cultivo de los predios agrícola de más de 5 hectáreas, según la nacionalidad de los propietarios, censo de 1930." *Ibid.* 5 (1): 3–4. Jan., 1942.

BUILDINGS, 1929

Dirección general de estadística.
Primer censo de edificios de los Estados Unidos Mexicanos. . . . México, D. F., Imprenta mundial, 1930. 119 pp. TH 28.A5 1930

BUILDINGS, 1939 [13]

Dirección general de estadística.
Segundo censo de edificios, 20 de octubre de 1939. Datos definitivos. México, D. F., Talleres gráficos de la nación. 1941. 32 vol. Govt. Publ. R. R.
There is a volume for each state, classifying all buildings according to use, material, number of rooms, number of dwelling units, type of ownership, sanitary facilities, etc. For some classifications, data are presented by county. Appropriate population data accompany the housing information.

EJIDOS, 1935

Dirección general de estadística.
Primer censo ejidal, 1935. México, D. F., Talleres gráficos de la nación, 1937. Govt. Publ. R. R.
The Library of Congress has received Volume I, Resumen general, and Volume II, numbers for Distrito Federal, Guanjuato, Hidalgo, Jalisco, México, Michoacán, Morelos, Puebla, Sonora, and Tlaxcala. The publication for Querétaro is in the library of the Pan American Union.
The characteristics of the *ejido* and its population are given according to the following classifications: total number of persons, payments made, number of persons in each ejido, type of organization, age of the association, and type and value of products.

EJIDOS, 1940

Dirección general de estadística.
Segundo censo ejidal de los Estados Unidos Mexicanos, 6 de marzo de 1940. Mexico, D. F., Publicaciones Secretaría de la economía nacional, 1942. Govt. Publ. R. R.

Aguascalientes. 133 pp.
Baja California—Territorio Norte. 104 pp.
Colima. 117 pp. (Dept. of Ag. Library).
Distrito Federal. 154 pp.
Quintana Roo. 119 pp.

INDUSTRY, 1930

Dirección general de estadística.
Primer censo industrial de 1930. México, D. F., Talleres gráficos de la nación, 1933–1935. 37 vol. HC 135.A5 1930

INDUSTRY, 1935

Dirección general de estadística.
Segundo censo industrial de 1935. México, D. F., Talleres gráficos de la nación, 1937–1941. Govt. Publ. R. R.
Separate volumes have been published for various industries.
Resumen general del censo industrial de 1935. México, D. F., Talleres gráficos de la nación, 1941. 250 pp. Pan Am. Union
The industrial labor force is classified by sex, industry, geographical units, industrial districts, and place of birth.

---

[13] Preliminary data from the 1939 census of buildings were published in Volumes III–V inclusive of the *Revista de estadística.* HC 131.R34

INDUSTRY, 1940

Dirección general de estadística.

Tercer censo industrial, 1940. *Revista de estadística 4 (6): 483–484.* Sept., 1941.

Tercer censo industrial, 1940. "Resumen de las características fundamentales de la industria." *Ibid.* 4 (7): 579–582. Oct., 1941. Govt. Publ. R. R.

CURRENT NATIONAL VITAL STATISTICS

(Including Population Estimates)

Departamento de salubridad pública. Dirección general de epidemiología.

Boletín de demografía y estadística sanitaria, Vol. 1, No. 1, junio de 1942. México, D. F., 1942. Govt. Publ. R. R.

This bulletin is to include reports on the demography of the states, municipalities and localities in which the Department of Public Health has offices, and on the public health activity in these various areas. The contents of the first issue are as follows: "El servicio de salubridad en la Republica Mexicana." Por Francisco de A. Benavides; "Geografía médica." Por Alfredo C. Aldama. "Geografía física y climatatología de la República." "Mortalidad por enfermedades infecciosas y parasitarias en la República." Por Ricardo Granillo. "Defunciones registrados en la ciudad de México, durante el año de 1940." Por Joaquín M. Sánchez.

Dirección general de estadística.

Anuario estadístico de los Estados Unidos Mexicanos, 1940. México, D. F., 1942. 806 pp. HA 762.A3

Section II. "Movimiento de la población," pp. 87–245, includes an introduction defining the concepts used in the classification of Mexican vital statistics. There is a summary table giving estimated total population and number and rates of marriages, births, deaths, and infant deaths, annually for 1893–1910 and 1922–1939. Current rates are based upon preliminary 1940 populations. Vital statistics for 1937–1940 are classified by rural-urban residence and by age, legitimacy, literacy, and occupation. Other tables present historical and contemporary information on migration and naturalization.

Mortalidad en México. México, D. F., 1942. 119 pp. Govt. Publ. R. R.

A graphic and tabular presentation of data for the nation and the various States.

Revista de estadística. México, D. F., March, 1938, to date. Govt. Publ. R. R.

National data are presented on marriages, divorces, births, deaths, infant deaths, deaths by cause, and natural increase. See: *Ibid,* 6(4). April, 1943. Since 1941 the inside back cover of each issue has contained series giving the numbers of and rates for births and deaths, 1930–1940. Special articles are sometimes included, i. e., "Defunciones por entidades y causas, año de 1940." *Ibid.* 5(2): 182–189. Feb., 1942.

Revista del Instituto de salubridad y enfermadades tropicales. México, D. F., 1940 to date. Govt. Publ. R. R.

Detailed studies in the general fields of mortality, morbidity and public health are often included in this periodical. For instance, see: Bustamante, Miguel E., and Aldama C., Alvaro. "Variación mensual del número de defunciones y principales causas de mortalidad por estados." *Ibid.* 2(3–4): 259–277. 1941. Also, by the same authors: "Mortalidad de menores de un año en la República Mexicana y en el Distrito Federal, 1922–1939." *Ibid.* 3(2): 81–92. 1942.

OTHER CURRENT OFFICIAL POPULATION STATISTICS

Bustamante, Miguel E., and Aldama C., Alvaro.

"Esperanza de vida" en veinte estados de la República Mexicana. *Revista del Instituto de salubridad y enfermedades tropicales 2(1): 5–18.* June, 1941. Govt. Publ. R. R.

Expectation of life at decennial ages is presented for twenty states comprising 74.71 percent of the population of Mexico and so chosen as to be representative of all geographical, climatological, and economic areas of the country. The provinces were Baja California (Territorio Norte), Sonora, Coahuila y Tamaulipas, Baja California (Territorio Sur), Sinaloa, Jalisco y Colima, Michoacán, Oaxaca y

Chiapas, Veracruz, Campeche y Yucatán, Zacatecas, Aguascalientes, Guanajuato, México, Puebla, and the Distrito Federal. The implications for public health and national policy are indicated.

Tablas de vida de los habitantes de los Estados Unidos Mexicanos. *Revista del Instituto de salubridad y enfermedades tropicales 1(2): 131–150.* May, 1940. Govt. Publ. R. R.

The provisional life tables, based on the 1930 census, are presented for the total population by sex, together with special tables for the State of Aguascalientes.

Departamento de salubridad pública.

Población y salubridad pública . . . por el Dr. Alberto P. León . . . Secretario general del Departamento de salubridad pública . . . México, D. F., 1940. 25 pp. HB 885.M4

The trends in population, natural increase, expectation of life, infant mortality, deaths from specific causes, and other indexes are compared for the United States and Mexico, with some reference to other American countries.

Dirección general de estadística.

Informes sobre las principales estadísticas Mexicanas. México, D. F., 1941. 174 pp. Govt. Publ. R. R.

This series of reports on the development and present status of the statistical activities of the Mexican government includes the following: "Censo de población," por Alfonso de Garay Alvarado; "Censos de edificios e industrias de la construcción," por Alfonso de Garay Alvarado; "Movimiento de población," por Manuel B. Trens Marentes; "Morbididad y sanidad," por Manuel B. Trens Marentes; "Censos ejidales," por Ing. Adolfo Alarcón Mendizabel.

México en cifras, 1938. México, D. F., Talleres gráficos de la nación, 1939. Unnumbered. HA 762.A5 19—

This illustrated compendium gives population data from the 1930 census, labor force data from the industrial census of 1935, vital statistics for 1936, and immigration and emigration for 1928–1936. Graphic materials on agricultural and industrial production, communication, commerce, and finance are included.

López Rosado, Diego G.

Atlas histórico geográfico de México. México, 1940. 109 pp. F 1227 L85

The section on the pre-Cortez epoch includes brief discussions and maps of the probable path of immigration into America, the major lines of the great migrations, the location and extension of the various cultural areas, principal cities and ethnic groups. The section on the colonial epochs includes a similar series of maps for the period from 1400 to 1786.

NICARAGUA

Historical

There is little definitive quantitative information on the historical development of the population of Nicaragua prior to 1920[1] The Guatemalan Province of Nicaragua was included in the Spanish count of 1778.[2] A trial census appears to have been taken in 1800; a new census, taken in 1813, was published in some detail in 1823 by General González Saravia in his *Bosquejo político, estadístico de Nicaragua.* This author also published estimates of the rate of natural increase, based on data relative to marital status and vital statistics compiled while he was Governor of the Province. Another census was attempted in 1834, but its results were recognized to be so defective that they were not published. A count is reported to have been attempted in 1845 and 1846. Another census was levied in 1867; materials available two decades later included a list of the inhabitants of each department, with name, age, sex, and occupation. However, these lists compiled by the *Ministerio de gobernación* appeared so grossly inadequate that the official gazette published figures raising the total population by almost four-fifths. A census in 1890 is cited by some students. There was some type of a census in 1906, although there are conflicting reports on its nature and extent. It has been variously described as a provisional census, a census of the department of Granada, and the only census to include business, livestock, and the mining industry. It was reputedly published in bulletins released regularly each month from July 1907 on.[3]

The value of these early estimates, partial counts, and incomplete censuses is seriously questioned by the students of Nicaraguan population history.[4] The first national census was taken in 1920, but even the introductory statement to the census itself discusses its inadequacies. Another national census was taken in 1940, but only preliminary data are available as yet.

The first *Anuario estadístico* of the *Dirección general de estadística,* published in 1939 and covering the year 1938, includes vital statistics. There is also a *Boletín mensual de estadística,* published currently, which gives quarterly or semiannual vital statistics. This serial has also served as the vehicle for the publication of the preliminary results of the 1940 population census.

---

[1] For general histories, see: Ayon, Tomás. *Historia de Nicaragua desde los tiempos más remotos hasta el año de 1852 . . .* Granada, Tipografía de El Centro Americano, 1882. 2 vol. F 1526.A98.

Gámez, José D. *Historia de Nicaragua desde los tiempos prehistóricos hasta 1860 . . .* Managua, Tipografía de El País, 1889. 855 pp. F 1526.G19.

Juarros, El Br. D. Domingo. *Compendio de la historia de la ciudad de Guatemala. . .* Tomo 1, 384 pp. 1808. Tomo 2, xv, 361 pp. 1818. F 1466.J89.

Oviedo y Valdés, Gonzalo Fernández de. *Histoire du Nicaragua.* Paris, A Bertrand, 1840. xv, 269 pp. (This is a translation of Chs. 1–13 of Book 42 of Oviedo's *Historia general de las Indias.*) vol. 14. E121.T32.

[2] For the history of the early estimates and counts, see: Gonzáles Saravia, Miguel. *Bosquejo político, estadístico de Nicaragua, formado en el año de 1823. . .* Guatemala, por Beteta, 1824. 1 pl. 23 pp. F 1526.G68.

Lévy, Pablo. *Notas geográficas y económicas sobre la República de Nicaragua. . .* Paris, Libería española de E. Denné Schmitz, 1873. 627 pp. F 1523.L66.

[3] Lindberg, Irving A. "Las actividades estadísticas de Nicaragua." pp. 381–384 in: Inter American Statistical Institute, *op. cit.*

[4] See, for instance: Squier, Ephraim G. *Nicaragua; its people, scenery, monuments, resources, condition, and proposed canal.* Rev. ed. New York, Harper and Bros., 1860. xvi, 691 pp. F 1523.S78.

EARLY ESTIMATES AND COUNTS

SECONDARY SOURCES

González Saravia, Miguel.
Bosquejo político, estadístico de Nicaragua, formado en el año de 1823 . . . Guatemala, por Beteta, 1824. 1 p. 1, 23 pp. F 1526.G68

Lévy, Pablo.
Notas geográficas y económicas sobre la República de Nicaragua . . . Paris, Librería española de E. Denné Schmitz, Comisionista para España y América, 1873. 627 pp. F 1523.L66

Lindberg, Irving A.
"Las actividades estadísticas de Nicaragua." Pp. 381–384 in: Inter American Statistical Institute, *op. cit.* Washington, 1941. xxxi, 842 pp. HA 175.S75

NATIONAL POPULATION CENSUSES

CENSUS OF 1920

Oficina central del censo.
Censo general de 1920. Managua, Tipografía nacional, 1920. xx, 327 pp. HA 831.A5 1920
The general and provincial sections give statistics on age, nationality, occupation, marital status, language, religion, sex, and color.

CENSUS OF 1940

Dirección general de estadística.
Boletín mensual de estadística. Números 20 a 25, julio a diciembre, 1940. Govt. Publ. R. R.
Preliminary figures on the population of provinces and capitals of provinces according to the 1940 census, p. 57–B.

CURRENT NATIONAL VITAL STATISTICS

Dirección general de estadística.
Boletín mensual de estadística. Números 20 a 25, julio a diciembre, 1940. Govt. Publ. R. R.
The section, "Demografía y migración," pp. 45–57–B, gives vital statistics for the third and fourth quarters of 1940 and a résumé for 1940, pp. 45–54. Deaths for departments are classified by marital status, occupations, age, and cause. Statistics on international migration are included.

Anuario estadístico general, 1938. Managua, D. N., 1939. 442 pp. HA 831.A3
"Sección demográfica," pp. 67–124, gives summary vital statistics for 1938, with detailed statistics on causes of death by departments. The following section, "Movimiento migratorio," pp. 125–170, gives statistics on international and internal movements, 1938.

Ministerio de la gobernación, justicia, policía, beneficencia y cultos.
Memoria de las Secretarías de gobernación y anexos y de beneficencia. Managua, Talleres nacionales, 1938–1939. 347 pp. J 183.R 25
The reports of the various municipalities often present vital statistics in summary form.

PANAMA

Historical

The general paucity of historical statistical materials for Panama is explainable in terms of the unsettled political history of the country. The Province of Panama was organized in 1719 under the Viceroyalty of Santa Fé. In 1739 it became a province in the Colombian federation of Nueva Granada. A union called the Isthmus of Panama was formed in 1841, but it was soon suppressed by Colombia. Between 1841 and 1903, when the independence of Panama was finally achieved, there were few periods of stability undisturbed by revolts. Even after the achievement of independence, administrative changes were so frequent as to prohibit the development of continuous statistical series of any types.[1]

The introductory statement to the first volume of the Panama census of 1920 discusses various early estimates, including one of 1793, but discards them as unreliable.[2] The territory was included in the Colombian census of 1870, but no further censuses were attempted prior to the national census of 1911.[3]

National censuses have been taken in 1911, 1920, 1930, and 1940. A law passed in 1906 provided for the taking of a census, but execution was delayed until 1911. The second census was taken in 1920; since only three of the provincial volumes could be located, there is a possibility that it was never published in full. The third national census, that of 1930, was the first in which the native people were enumerated rather than estimated. The fourth national census was taken in 1940. Careful preparations were made, with tracts delimited in such a way that the enumerator could cover his territory in 8 hours. Only questions on sex, knowledge of Spanish and literacy were asked of the native peoples. Schedules were edited before the release of the preliminary figures, and when necessary were returned for completion. The preliminary releases are now appearing in *Estadística panameña,* the new serial publication of the *Dirección general de estadística.*

The administrative changes and the general poverty of the governmental offices have reacted against the possibility of continuous collection and publication of vital statistics. The *Anuario de estadística,* 1934, issued by the *Secretaría de agricultura y obras públicas,* was the first yearbook in 23 years, although sporadic publications between 1903 and 1934 had given vital statistics. The *Sección de estadística* of the *Ministerio de agricultura y comercio* issued an *Anuario* in 1939 which contained demographic statistics. Current vital rates are published occasionally in the *Boletín de trabajo, comercio e industrias,* as well as in the new *Estadística panameña.*

---

1 Peralta, Manuel María de. *Costa Rica, Nicaragua y Panamá en el siglo xvi; su historia y sus límites según los documentos del Archivo de Indias de Sevilla . . .* Madrid, M. Murillo, 1883. xxii, 832 pp. F1437.P42

See also· Bell, Eleanor Y. "The Republic of Panama and its people, with special reference to the Indians." pp. 607–637 in: *Smithsonian Institution, Annual Report, 1909.* Washington, 1910. Q 11.S66 1909

2 Dirección general del censo. *Boletín número 1,* p. 3. Panamá, Imprenta nacional, 1922. HA 857.A38

3 Müller, Hans J. "Las actividades estadísticas de Panamá." Pp. 392–434 in: InterAmerican Statistical Institute, *op. cit.* See also: *Economic Literature of Latin America.* Ibid., pp. 791–792.

NATIONAL CENSUSES

EARLY CENSUSES AND ESTIMATES

For nineteenth century censuses and estimates, *see* Colombia.

CENSUS OF 1911

Laws, statutes, etc.

Primer censo de población en la República de Panamá. Leyes y decretos sobre censo de 1910. Ed. oficial. Panamá, Escuela de artes y oficios, 1910. 8 pp. HA 852.A5 1910

Dirección general de estadística.

Boletín del censo de la República de Panamá. Panamá, Imprenta nacional, 1911. vii, 18 pp. HA 851.A4 1911

Data from the Census of 1911.

CENSUS OF 1920

Dirección general del censo.

Boletín número 1. *Censo demográfico de la Provincia de Panamá, 1920.* Panamá, Imprenta nacional, 1922. 260 pp.

Boletín número 2. *Censo demográfico de la Provincia de Colón, 1920.* Panamá, Imprenta nacional, 1922. 158 pp.

Boletín número 3. *Censo demográfico de la Provincia de Cocle, 1920.* Panamá, Imprenta nacional, 1922. 126 pp. HA 851.A38

CENSUS OF 1930

Dirección general del censo.

1930 censo demográfico. Panamá, Imprenta nacional, 1931–1932. 2 vol. HA 851.A4

Subjects covered include size, distribution, sex, education, physical defects, marital status, religion, race, and nationality, age, place of birth, and occupation. The first volume includes the summary for the Republic and tables for provinces of Bocas del Toro, Cocle, Colón, Chiriqui, Darien, and Herbera. The second volume continues with the provinces of Los Santos, Panamá, and Veraguas. The remainder of the volume contains lists of places of over 750 inhabitants.

CENSUS OF 1940

Contraloría general de la República. Oficina del censo.

Censo de población, 1940. Informe preliminar. Panamá, Imprenta nacional, 1943. 36 pp. Bur. of Cen.

Tables: (1) Population by provinces, censuses of 1911, 1920, 1930 and 1940. (2) Density, by provinces, same dates. (3) Population of urban centers, 1930 and 1940. (4) Civilian population, by provinces and minor civil divisions, by sex, 1940. (5) Civilian and native populations, provinces. (6) Population of the city of Panama, by barrio, by race and sex. (7) Population of the city of Colon. *Ibid.*

The history of the 1940 census is sketched. Final publications will include nine provincial volumes, with a section in each covering geography, history, natural resources, political divisions.

Contraloría general de la República. Sección de estadística.

Demografía. *Ibid.* 1 (8): 1–2. May, 1942. [Population ten years of age and over, by groups of age, literacy, etc., City of Panama, Census of 1940.] Govt. Publ. R. R.

Población económicamente activa de 10 años o más, por grupos de edades y sexo. Ciudad de Panamá: Censo de 1940. *Ibid.* 1 (9): 1 June 1942. [The succeeding pages give tables on the gainfully occupied population ten years of age and over: (1) By type of activity and by sex, City of Panama. (2) By age groups and sex, City of Colon. (3) By type of activity and sex, City of Colon.]

Población nacida en el extranjero, por sexo y raza, Ciudad de Panamá, Censo de 1940; Población nacida en el extranjero, por nacionalidad y sexo, Ciudad de Panamá, Censo de 1940; Población nacida en el extranjero, por sexo y raza, Ciudad de Colón, Censo de 1940; Población nacida en el extranjero, por nacionalidad y sexo, Ciudad de Colón, Censo de 1940. *Ibid* 1 (10): 1–4. July, 1942.

Población nacida en la República de Panamá, por sexo y raza, Ciudad de Panamá, Censo de 1940; Población nacida en la República de Panamá, por sexo y raza, Ciudad de Colón, Censo de 1940; Población nacida en la República de Panamá, por provincia de nacimiento y sexo, Ciudad de Panamá, Censo de 1940. *Ibid.* 1 (11): 1–3. Aug., 1942.

Población activa de la Ciudad de Panamá que trabajaba en la Zona del Canal, por ocupación y sexo, Censo de 1940. *Ibid.* 2 (1): 4–5. Oct., 1942.

Número de personas de 10 años o más que trabajan en el comercio, por clase de negocio y sexo, Ciudad de Panamá, Censo de 1940. *Ibid.* 2 (2): 4. Nov., 1942.

OTHER NATIONAL CENSUSES

Secretaría de trabajo, comercio e industrias.

Análisis de los resultados del primer censo oficial de empresas industriales y comerciales radicadas en Panamá, ejecutado en el año de 1938. *Boletín de trabajo, comercio e industrias (13): 2–26,* 1939. Govt. Publ. R. R.

The analysis of the 1938 census, which covered the entire Republic, includes the geographical distribution of industry by type. There are data on citizenship status of proprietors by various characteristics of their business and nationality by industry.

CURRENT NATIONAL VITAL STATISTICS

Contraloría general de la República. Sección de estadística.

Estadística panameña 2 (1), Oct., 1942, and 2 (2), Nov., 1942. Govt. Publ. R. R.

Statistics on causes of deaths by age and sex are given for the cities of Panama and Colon, 1941.

Dirección general de estadística.

Anuario de estadística, año 1934. No. 82. Panamá, Imprenta nacional, 1936. 418 pp. HA 851.A35

The section, "Demografía," pp. 3–36, reports migration, births, marriages, and deaths as of 1934. Births, marriages, infant mortality, general mortality, and divorces are reported in detail with distributions according to minor divisions, sex, marital status, cause and age.

Data for succeeding years were assembled, but the *Anuario* was not published. The *Ministerio de agricultura y comercio, sección de estadística,* issued a new *Anuario* for 1939 in 1940, but no copy was located.

Secretaría de trabajo, comercio e industrias.

Boletín de trabajo, comercio e industrias. No. 2, May, 1938, to No. 28, July, 1940. Govt. Publ. R. R.

Demographic statistics were currently published in this bulletin.

See: "Estadística de natalidad y mortalidad desde el año de 1931 al de 1937. *Ibid.* (7): 11. Oct., 1938.

Movimiento de población. Movimiento demográfico registrado en la República de Panamá, según datos suministrados por los alcaldes a la Sección de estadística, durante el mes de agosto de 1939. *Ibid.* (17): 3. Aug., 1939.

Movimiento de población. Movimiento demográfico registrado por los alcaldes durante el segundo trimestre del año de 1940. *Ibid.* (28): 16–17. July, 1940.

"Cuadro demostrativo del movimiento internacional de pasajeros en la República de Panamá durante el año de 1939." *Ibid.* (23): 17. Feb., 1940.

"Movimiento de pasajeros por los puertos de la República durante el mes de junio de 1940." *Ibid.* (29): 9. Sept., 1940.

OTHER OFFICIAL PUBLICATIONS

Laws, Statutes, etc.

Nueva división territorial de la República de Panamá, ley número 103, con un índice y una nomenclatura. Panamá, Imprenta nacional, 1941. 62 pp. Govt. Publ. R. R.

PARAGUAY

Historical

The economic and political history of Paraguay has been such that no complete and accurate census has ever been taken, although counts have been attempted several times in the nation's history. A long series of estimates lay back of the census of 1886. The first issue of the *Anuario estadístico*, that for 1886, presents estimates for the year 1536, 1775, 1828, 1852, 1857, 1861, 1872, and 1886.[1] According to these estimates, the population increased from less than 100,000 in 1536 to 1,300,000 in 1861, and then was reduced to 231,000 in 1886 after the War of 1865–70 against Argentina, Brazil and Uruguay. In the absence of any information as to the basis of the estimates, these figures must be accepted with considerable reservation. There is little doubt but that the losses in the war against the Triple Alliance were great, but the difference between the estimate of 1861 and the census population of 231,000 in 1886 cannot be assumed to constitute a measure of their magnitude.[2]

The first national census was taken on March 1, 1886, in order to determine the size of the legislature. The probable extent of its incompleteness is indicated by the fact that its results were not accepted by the officials. A special section of the 1886 *Anuario*, entitled *Cuadros del censo general de la República*, presents tables on population by sex and nationality, and some age data for 37 parts of the country.[3] A census was taken during the years 1899–1900, but no publications or official references were discovered.[4] An electoral census was taken in 1917. An attempt was made to secure authorization for a census in 1930, but it failed.[5] A population census was attempted in 1936-37, although the revolution at the beginning of 1936 impeded the work.[6] There seems to be considerable question within Paraguay as to the completeness of this census. Significantly, no published data from it could be located. The Director General of Statistics, Alfonso B. Campos, reported in 1940 that the total population of the country for this "census" was consistent with that estimated on the basis of the vital rates of the city of Asunción.[7]

Vital statistics based on a Civil Registry system are published currently, but detailed tabulations are available primarily for the city of Asunción.

NATIONAL POPULATION CENSUSES

CENSUS OF 1886

Oficina general de estadística.

Anuario estadístico de la República del Paraguay, año 1886. Libro primero del anuario. Asunción, Fischer y Quell, 1888. viii, 275 pp. HA 1041.A2

[1] Oficina general de estadística. *Anuario estadístico de la República del Paraguay, año 1886*. . . Asunción Fischer y Quell, 1888. viii, 275 pp. See especially Ch. II, pp. 41–64. HA 1041.A2

[2] James estimates the population in 1865 as 525,000, that in 1870 as 300,000, of whom only 22,000 were males. See pp. 273–274 in: James, Preston E. *Latin America.* New York, Odyssey Press, 1942. xviii,906 pp. For a history of the war, see pp. 210–251 in: Schuster, Adolph N., *Paraguay.* Land, Volk, Geschichte, Wirtschaftsleben und Kolonisation. Stuttgart, Strecker und Schröder, 1929. xx, 667 pp. F. 2668.S39.

[3] Paraguay—Oficina general de estadística. *Op. cit.*, pp. 212–265.

[4] Inter American Statistical Institute. *Op cit.*, p. 439. HA 175.S75.

The issues of the *Anuario estadístico* for 1914–1917 inclusive contain neither population estimates nor references to censuses. See: Paraguay. Dirección general de estadística. *Anuario estadístico de la República del Paraguay, 1914.* Asunción, Talleres gráficos del Estado, 1914. 12 parts, numbered separately. HA 1041.A2.

[5] *Economic Literature of Latin America. Op. cit.*, p. 765.

[6] Campos, Alfonso B. "Las actividades estadísticas del Paraguay." Pp. 440–442 in: Inter American Statistical Institute. *Op. cit.*

[7] ". . . la cifra de la población del país hallada por medio de dicho censo, corresponde al porcentaje de natalidad del movimiento demográfico de la capital determinado sobre cifras positivas." *Ibid.*, p. 442.

The introductory section on the history of Paraguay presents a graph, "Cuadro comparativo del aumento de la población, desde el año 1536 hasta el año 1886," apparently based on estimates made in 1536, 1775, 1828, 1852, 1857, 1861, 1872, and 1886. According to this, the population, less than 100,000 in 1536, increased to 1,300,000 in 1861, and was reduced to about 231,000 in 1886.

Ch. II, pp. 41–64, summarizes early estimates of population. It, together with the final section of the yearbook, pp. 212–265, summarizes the results of the census of March 1, 1886, with reference to population by sex, nationality, citizenship status, physical defects, education, distribution and occupations. Data are given for 37 parts of the country. Vital statistics are included.

ELECTORAL CENSUS OF 1917

Dirección general de estadística.

Censo electoral. Elecciones ordinarias de senadores y diputados verificadas el 4 de marzo de 1917. Asunción, Talleres gráficos del Estado, 1917. 44 pp.
Pan Am. Union

CURRENT NATIONAL AND CITY VITAL STATISTICS

(Including Population Estimates)

Dirección general de estadística.

Boletín semestral. Segundo semestre de 1937. Vol. 23, Nos. 91 and 92. Asunción, 1937. 49 pp. HC 223.A8A2

Sección II, "Población," gives data on immigration, births, marriages and deaths for the city of Asunción. Nos. 101–102 and 103–104 for 1940 are now available.

Dirección general de estadística.

Memoria . . . correspondiente al año 1938. Asunción, Imprenta nacional, 1941. 246 pp. HA 1041.A35

Sección III, "Demografía," pp. 29–38, presents vital rates for Asunción, 1932–1938; migration 1932–1938; deaths by causes for Asunción, 1937 and 1938; immigrants by nationality, occupation, sex and marital status, 1938.

The *memoria* for 1939 was published in 1942.

Ministerio de economía.

Anuario del ministerio de economía, 1938–39. Asunción, 1939. 546 pp.
HC 221.A25

A brief section, "Población del Paraguay," pp. 21–25, gives population estimates for the capital and the departments as of Dec. 1937, together with estimates of racial composition and vital rates for the country as a whole. Vital statistics are given for the city of Asunción, 1938.

Ministerio de salud pública.

Boletín . . . No. I. Agosto, 1940. Asunción, La Colmena, [1940?] 192 pp.
Govt. Publ. R. R.

The report of the *Departamento de higiene, Oficina central y con registradores locales,* pp. 87–98, contains a brief description of the organization. Tables give data for the first half of 1940 for cities and pueblos on births by sex and legitimacy status; marriages, by nationality; morbidity by causes by locations; deaths in Asunción and in the 24 pueblos; deaths due to infections and parasitic diseases, Asunción; causes of death without medical assistance in Asunción and 24 pueblos combined; causes of death by captions of the International Nomenclature, by localities; infant mortality by months by localities. The following section presents epidemiological statistics.

Memoria de la Sección estadística . . . 1938. Asunción, Imp. en los talleres de T. Navarro, Azara esq. Yegros, 1939. 131 pp. HA 475.A3

Statistics on births, deaths, infant mortality, and causes of death are presented for 25 cities and pueblos for 1938, pp. 9–42.

The report for 1939 was published in 1942.

OTHER OFFICIAL PUBLICATIONS

Ministerio de economía.

Paraguay, datos y cifras estadísticas: población, producción, importación, exportación, industrias, vialidad, comercio, instrucción pública. Publicación oficial del Ministerio de economía. Asunción, 1939. 30 pp., tables. HA 1044.A5 1939

PERU

Historical

The first quantitative data on the size and distribution of the population of Peru were secured by the Inca. The centralized political organization and economy of their Empire necessitated the use of estimates or statistics on the number and characteristics of the population. Counting was done by means of quipu, cords of multicolored threads in which knots were used to represent fixed numbers. A centralized administrative system was developed to transmit the data from the local areas to the capital city.[1]

Population and other statistical data for the period of the Viceroyalty are included in the volumes of the *Mercurio Peruano*, published in Lima between 1791 and 1795.[2] Many estimates, partial enumerations, and even complete local enumerations were made throughout the seventeenth and eighteenth centuries. The most detailed information was that secured in the censuses of 1741 and 1795, both of which were taken by order of the King of Spain.[3]

There were numerous statistical compilations of population data during the early part of the nineteenth century. The Constitutions of 1823 and 1828 had given the departmental councils authority to take population censuses; the Constitution of 1855 gave similar powers to the municipalities. In 1828 the municipalities were ordered to make civil registers of the citizens having the right to vote; in 1834 the order was extended to include the making of a complete enumeration of the population. The *Consejo central directiva* was established in 1848, with the responsibility of developing and coordinating statistics of all types. However, this organization did not prove particularly successful, and a law of 1853 provided for the establishment of a *Sección estadística* in the *Ministerio de gobierno*, which was to assume responsibility for census enumeration. A national population census was taken in 1862.

Further reorganization of the statistical service occurred in 1873, when the *Dirección de estadística* was divided into three sections, *Estadística de población*, *Estadística del territorio*, and *Estadística del estado*. A vital statistics service was organized in 1876, and a national population census was taken the same year.

---

[1] Kürbs, Friedrich. "Las actividades estadísticas del Perú." pp. 451–461 : Inter American Statistical Institute. *op. cit.* See also: Arco Parró, Alberto. "Problemas y soluciones para el censo demográfico Peruano de 1940." pp. 17-27 in: *Eighth American Scientific Congress. Proceedings, Vol. VIII.* Statistics. Washington, Dept. of State, 1942. 365 pp.

[2] *Mercurio peruano de historia, literatura, y noticias públicas que da á luz la Sociedad académica de Amantes de Lima.* . . . Lima, Imprenta real de los niños huérfanos, 1791–1795. 12 vol. F 3401.M55

Historical studies were also included; e. g., Reflexiones históricas y políticas sobre el estado de la población de esta capital, que se acompaña por suplemento. . . . Serie de las recensiones, con el aumento o diminución que ha habido de una á otra según lo demostrado. 3 de febrero de 1791, pp. 91–97.

[3] For the history of censuses and studies of the trend of population growth in Peru, see especially: Paz Soldán, Mariano Felipe. *Diccionario geográfico estadístico del Perú, contiene además la etimología aymara y quechua de las principales publicaciones, lagos, ríos, cerros* . . . Lima, Imprenta del estado, 1877. 1077 pp. F 3404.P34.

See also: Capelo, Joaquín. *La despoblación, Perú.* Casa editoral, Sanmartí y Cía., 1912. 20 pp. HB 3575.C3

And: Graña y Reyes, Francisco. *La poblacion del Perú a través de la historia.* Tercera edición. Lima, Imprenta Torres Aguirre, 1940. 50 pp. HB 3575.G7 1916c

This census, issued in seven volumes, is not only the first comprehensive census published for the country, but also the only national census in the 64-year period between 1876 and 1940. Great difficulties were experienced in the actual enumeration, partially because of the complexity of the census itself, partially because of the general conditions of the country at that time. A law providing for personal taxation for school purposes had just been passed, and large proportions of the people assumed that the census was intended as a base for the preparation of tax lists.

The estimation of the size of the population of Peru during the several decades before the census of 1940 was extremely difficult. There were some censuses of special groups and special areas, but no enumeration of the total population.[4] Hence the data of the census of 1940 were of great importance to all the statistical organizations of Peru. Careful preparations were made to insure its completeness and accuracy. Preliminary surveys were carried out to establish exact lists of populated centers, a careful tract delimitation procedure was followed, and an intensive educational campaign was carried on through posters, direct mail, and the school system. Enumerators were carefully chosen and instructed. Special care was taken in connection with remote places, such as those of very high altitudes. In the case of the Amazon and other jungle Indians, reliance had to be placed on the estimates of missionaries, traders, and governmental personnel who were in intimate contact with the Indians.

Different types of schedules were used for the various size classes of the population, the amount of the data requested being greatest for the nine largest cities and least in the rural areas. The census schedules included questions on sex, race, age, marital status, education, religion, nationality, and occupation. Fertility data, apparently collected only for persons enumerated on the so-called family schedule, i. e., those in the nine largest cities, included the number of children born alive, number of years since first birth, and number of children still living.

The annual publication of the *Dirección nacional de estadística*, the *Extracto estadístico del Perú*, is the most accessible source of population data for Peru. Vital statistics for all of Peru have been published in this annual since 1923. A monthly *Boletín*, also issued by the *Dirección nacional de estadística*, presents some demographic data, but only two issues, one of 1936 and one of 1938, were located. The *Dirección general de salubridad* issues a quarterly, the *Boletín del Instituto nacional del niño*. Periodicals published by some municipalities, such as Lima and Cuzco present vital statistics in greater detail or of a more recent date than those in the *Extracto estadístico.*

NATIONAL POPULATION CENSUSES

EARLY CENSUSES AND ESTIMATES

Kürbs, Friedrich.
"Las actividades estadísticas del Perú." pp. 451–461 in: Inter American Statistical Institute, op. cit. Washington, 1941. xxxi, 842 pp. HA 175.S75

. . *Mercurio peruano de historia, literatura y noticias públicas que da á luz la Sociedad académica de amantes de Lima.* . . . Lima, Imprenta real de los niños huérfanos, 1791–1795. 12 vol. F 3401.M55

---

[4] An educational census of the population 4 to 14 years of age was taken in 1902 in order to determine the primary school capacity needed. There was also an electoral census in 1933, published as: Servicio de estadística y censo electoral. *Extracto estadístico y censo electoral de la República.* Lima, 1933. HA 1055.A5 1923.

Many histories of Peruvian statistics report a census of 1896, but this was actually an estimate made by a commission presided over by Vice-Almirante don M. Meliton Carbajal, vice-president of the *Sociedad geográfica de Lima.* See: Dirección nacional de estadística. *Extracto estadístico del Perú, 1938.* Lima, Imprenta Americana, 1939. xlvi, 644 pp. Table I, pp. 12–14.

Paz Soldán, Mariano Felipe.
Diccionario geográfico estadístico del Perú, contiene además la etimología aymara y quecha de las principales poblaciones, lagos, ríos, cerros, etc. Lima, Imprenta del Estado, 1877. xxix, 1077 pp. F 3404.P34

CENSUS OF 1862

The original report of this census was not located.

CENSUS OF 1876

Dirección general de estadística.
Censo general de la República del Perú formado en 1876. Lima, Imprenta del teatro, 1878. 7 vol. Univ. of Chicago Library
Resumen del censo general de habitantes del Perú hecho en 1876. Lima, Imprenta del estado, 1878. 853 pp. HA 1052.A5 1876

CENSUS OF 1940

Departamento de censos.
Censo nacional de 1940. Resultados generales. Primer informe oficial. Lima, 1941. 68 pp. HA 1051.A55
Contents: Introduction and Chs. I–V. Census decrees and plans; organization and execution of the census; plans of analysis and publication. Ch. VI: General results. Ch. VII: Estimates of omissions. Ch. VIII: Summary of population increase. Ch. IX: Analytical study of the composition of the population: age 0–5, 6–14, 15–19, 20–59, and 60 and over; age, sex, and race for departments. Ch. X: The census and the control of national population statistics.

Dirección nacional de estadística.
Estado de la instrucción en el Perú según el censo nacional de 1940. Informe especial. Lima, 1942. 67 pp. Govt. Publ. R. R.
The population of school age is given by regions.

OTHER NATIONAL CENSUSES

CENSUS OF EDUCATION, 1902

Dirección de primera enseñanza.
Censo escolar de la República Peruana correspondiente al año 1902. Lima, Imprenta Torres Aguirre, 1903. 582 pp. L 324.B5 1902
Persons aged 4 to 14 only. Information by small areas. Age, race, literacy, whether in school, whether completed required schooling, citizenship. Incomplete and partial.

PROVINCIAL CENSUSES

Junta departamental de Lima pro-desocupados.
Censo de las provincias de Lima y Callao levantado el 13 de noviembre de 1931. Lima, Imprenta Torres Aguirre, 1931. HA 1068.L7A5 1931
Detailed occupational data, pp. 192–247; unemployed persons by age, nationality, etc., pp. 248–257; number of children ever born, for married, widowed, and divorced women, and year of marriage for childless married or widowed women, pp. 260–265.

CURRENT NATIONAL VITAL STATISTICS

Dirección nacional de estadística.
Extracto estadístico del Perú, 1940. Lima, Imprenta Americana, 1941. Annual. HA 1052.A4
The section on *Movimiento de la población*, pp. 62–140, 1940, includes the following materials:
Data for the evaluation of vital statistics: number of districts reporting births, deaths, and marriages in 1940 (number, and as percent of those which should have reported), monthly, and annual average; vital statistics schedules (number received compared with and expressed as percent of number which should have been received), by departments, separately for births, deaths, and marriages.
Summary birth, death, and marriage statistics tables for Peru as a whole, annually 1923–1940; same information for departments monthly, by districts, and by occupation of parents, 1940.

Marriages cross-classified by age of parties, race, and nationality; and for various combinations of citizens and noncitizens, by months, and for selected combinations of previous marital status.

Births: stillbirths and live births by race and legitimacy, cross-classified by age of parents; multiple birth by sex, legitimacy, and race; births cross-classified by nationality of parents. Total country only.

Deaths by age, race, and sex; by age, nationality, and marital condition; and for the seventeen principal causes of death by sex, for departments.

Dirección nacional de estadística.

Boletín. Lima, Imprenta Americana. Monthly. 19—. HA1051.A3

The section, "Demografía," presents general population and vital statistics. Only two issues were located, one for 1936 and one for 1938.

Dirección general de salubridad.

Boletín del instituto nacional del niño. Lima. Quarterly, 1941. Govt. Publ. R. R.

This publication occasionally contains analyses of vital statistics. For instance, the issue for July–September, 1941, contains a study of infant mortality in Lima.

CITY VITAL STATISTICS

Inspección de estadística y demografía del Concejo provincial de Lima.

Boletín demográfico municipal de la ciudad de Lima. Vol. VII, No. 32, Oct.–Dec., 1939. Govt. Publ. R. R.

Detailed vital statistics are given for the city of Lima.

Concejo provincial del Cuzco.

Boletín municipal. Cuzco, 19—. Formerly monthly, now bi-monthly Govt. Publ. R. R.

Detailed vital statistics are included. The last issue available is that for Nov.–Dec., 1941.

UNITED STATES

Historical

The local and colonial governments in the area which later became the United States made many attempts to estimate or count the numbers of their citizens. Knowledge of the size of the population was desired primarily because it would permit estimates of tax yields and of manpower available for defense against Indian or other attacks. The inquiries of the Lords of Trade concerning population were also an important factor stimulating many governors to undertake censuses.[1] Most of the population data from the Colonial period, however, consist of partial enumerations and estimates of various types. The latter were usually computed on the basis of the number of militia, polls, tax lists, and families, or houses, with a multiplier used to secure total population. A recent attempt to determine the population of the Colonies in 1775 had to rely on partial estimates for New Jersey, Delaware, and Virginia, on tax lists for Pennsylvania and North Carolina, and on governors' estimates and land grants for Georgia.[2]

The Constitution of the United States provided that an enumeration of the population "be made within 3 years after the first meeting of the Congress of the United States and within every subsequent term of 10 years." [3] The first census, that of 1790, was taken by United States marshals, who secured information on the name of the head of the family, and the number of white males under 16 years old and 16 and over, white females, other free persons, and slaves. The censuses between 1790 and 1900 continued to be taken by an organization created temporarily for the purpose, although the scope of the census inquiries and the complexity of the tabulations expanded fairly continuously through the ensuing century of rapid economic and political expansion. An industrial census was first attempted in 1810, although the results were not particularly satisfactory. The scope of the census was broadened in 1840 to cover population, mines, agriculture, commerce, manufactures, and schools, with its defined objective to " . . . exhibit a full view of the pursuits, industry, education and resources of the country."

Technical improvements accompanied the expansion of the scope of the census. In 1830 a printed schedule was used for the first time. The census of 1850 was the first to be taken on the basis of a separate line for each individual. In 1880 a separate and specially trained force of supervisors and enumerators replaced the United States marshals as enumerators, and an attempt was made to obtain the

[1] For histories of these early censuses and estimates, see: Dexter, Francis B. "Estimates of population in the American Colonies." *Proceedings*, American Antiquarian Society, N. S. 5: 22-50. 1889.
E 172.A357-A358

Greene, Evarts B., and Harrington, Virginia D. *American population before the Federal Census of 1790.* New York, Columbia University Press, 1932. xxii, 228 pp. HB 3505.G 7.

Sutherland, Stella H. *Population distribution in Colonial America.* New York, Columbia University Press, 1936. xxxii, 353 pp. HB 1965.S84

[2] Sutherland, Stella H., *op. cit.*

[3] This résumé of the history of the decennial census is based on a more detailed historical note prepared by the Division of Statistical Research of the Bureau of the Census.

data in one month instead of the 10 to 20 required for previous censuses. Tabulations by electrical machinery appeared in connection with the Census of 1890. Finally, in 1902, a permanent Census Bureau was established in the Department of the Interior, later transferred to the Department of Commerce and Labor, and finally to the Department of Commerce.

The technical and administrative improvements between the first scientific census, that of 1850, and the comprehensive sixteenth census of 1940, have permitted a vast expansion in the amount and variety of data secured. The population census itself has expanded to a comprehensive report on such personal characteristics as age, color, sex, marital status, and place of birth, education, migration, employment status, occupation, and income; and for each household, tenure and value or rent. In 1940 a 5-percent sample of the population also provided information on nativity of parents, mother tongue, veteran's status, social security, usual occupation, industry and class of worker and, if a married, widowed, or divorced woman, the number of children ever born. A census of housing was included for the first time in 1940.

Censuses of manufacturing were taken in 1810, 1820, and 1840, decennially from then to 1904, quinquennially from 1904 to 1919, and biennially from 1919 until the suspension of the 1941 enumeration because of the war. A special census of mineral industries has been taken decennially since 1840. Information on agriculture has been gathered at every decennial census since 1840, and in quinquennial censuses of agriculture in 1925 and 1935. Special surveys of irrigation and drainage have been made in conjunction with the census of agriculture, the former since 1910, the latter since 1920. Censuses of business were taken in 1930, 1933, 1935, and 1939, covering retail trade, wholesale trade, and various classes of service establishments. Construction was included in 1929, 1935, and 1939. Information on churches was obtained in the decennial censuses for some decades after 1850. In 1906, this subject became a separate inquiry instead of being included in the decennial census as it had been until 1900. Censuses of religious bodies are now taken decennially in every year ending in six.

The first official vital statistics for the United States were collected as a part of the decennial census of 1850. Some vital data were requested in each census between 1850 and 1900, but when the Bureau of the Census was made permanent in 1902 the Director was authorized to obtain copies of records filed in the vital statistics offices of those states and cities which had registration systems meeting the minimum standards set by the Director. The annual collection of mortality statistics began with the calendar year 1900, although a national birth registration area was not established until 1915. Birth and death registration areas did not become nation-wide until 1933.

It is obvious that a complete bibliography of the census publications of the United States could not have been included within the limited space of this bibliography, even if personnel had been available for the gigantic task involved in the compilation of such a list. Hence the bibliography which follows is limited to the major published volumes of the Decennial Censuses and current national vital statistics. More detailed historical and bibliographical surveys are included in the following publications:

Bureau of Labor Statistics.

The history and growth of the United States Census. Prepared for the Senate Committee on the Census, by Carroll D. Wright, assisted by William C. Hunt . . . Washington, Govt. Printing Office, 1900. 967 pp.

HA 37.U5 1900

Bureau of the Census.

A century of population growth from the first census of the United States to the twelfth, 1790–1900. Washington, Govt. Printing Office, 1909. x, 303 pp.

HA 195.A5

Bureau of the Census.
Circular of information concerning census publications, 1790–1916. January 1, 1917 . . . Washington, Govt. Printing Office, 1917. 124 pp.
Z 7554.U5U5 1916
Z 1223.C4 1916

Superintendent of Documents.
Census publications; statistics of population, agriculture, manufactures, and mining, with abstracts and compendiums; list of publications relative to above subjects for sale by the Superintendent of Documents, Govt. Printing Office, Washington, D. C. . . . Washington, 1920. Z 1223.A191 No. 70
Z 7554.U5U8

Bureau of the Census.
Index of data tabulated from the 1930 census of population including unemployment. . . . Washington, Govt. Printing Office, 1940. 47 pp.
HA 205.A5 1930 h

Bureau of the Census.
Key to the published and tabulated data for small areas (preliminary). Population, housing, business, manufactures, agriculture. Washington, 1942. 66 pp.

Bureau of the Census.
Census Bureau publications. April 15, 1943. 18 pp.
Publications of the Bureau available from the Superintendent of Documents are listed separately for the fields of Population and Housing, Business, Current Manufactures Reports, Basic Materials, Vital Statistics, State and Local Government, Field Service, Statistical Research, and Agriculture. This list is issued monthly.

Department of Commerce.
Thirteenth annual report of the Secretary of Commerce, 1942. Washington, Govt. Printing Office, 1942. 136 pp. HF 105.C23
The annual report of the Director of the Bureau of the Census is included, pp. 15–79. These annual reports of the Director constitute invaluable source material for the study of the year-to-year activities of the Bureau of the Census, as well as the progress of the current decennial censuses.

DECENNIAL CENSUSES

CENSUS OF 1790

General

Census Office. 1st Census, 1790.
Return of the whole number of persons within the several districts of the United States . . . Printed by order of the House of Representatives. Philadelphia, printed by Joseph Gales, 1791? 56 pp. HA 201.1790

Other

Bureau of the Census.
Heads of families at the First Census of the United States taken in the year 1790 . . . Washington, Govt. Print. Office, 1907–08. 12 vol. HA 201.1790.C
The records of the names of the heads of the families in 1790 are reproduced for Maine, New Hampshire, Vermont, Massachusetts, Rhode Island, Connecticut, New York, Pennsylvania, Maryland, Virginia, North Carolina, and South Carolina.

CENSUS OF 1800

General

Census Office. 2d Census, 1800.
Return of the whole number of persons within the several districts of the United States, . . . Printed by order of the House of Representatives. Washington, 1801. 3 pp. l., [3]–34, [36]pp. HA 201. 1800.A

CENSUS OF 1810

General

Census Office. 3d Census, 1810.
Aggregate amount of each description of persons within the United States of America, and the territories thereof, agreeably to actual enumeration made according to law, in the year 1810. Washington, 1811. 90 pp. HA 201.1810.B

Manufacturing

Treasury Dept.

A statement of the arts and manufactures of the United States of America, for the year 1810 . . . Philadelphia, Printed by A. Cornman, Junr., 1814. 6 pp. l., v–[lxiv]pp. (p. lxiv incorrectly numbered xiv), 1 p. l., 46, 169 (i. e. 171) pp. HA 201.1810.B2

Tabular statements of the several branches of American manufactures, exhibiting them: (I) By States, territories and districts, (II) By counties, cities, and towns, so far as they are returned in the reports of the marshals, and of the secretaries of the territories, and their respective assistants, in the autumn of the year 1810; together with similar returns of certain doubtful goods, productions of the soil and agricultural stock, as far as they have been received. Philadelphia. Printed by A. Cornman, Junr., 1813. 1 p. l., 46, 169 (i. e. 171)pp. HD 9724.A3

CENSUS OF 1820

General

Census Office. 4th Census, 1820.

Census for 1820. Published by authority of an act of Congress, under the direction of the Secretary of State. Washington, Printed by Gales & Seaton, 1821. 80 pp. HA 201.1820.B1–1

Manufacturing

Digest of accounts of manufacturing establishments in the United States, and of their manufactures. Made under direction of the Secretary of State, in pursuance of a resolution of Congress of 30th March, 1822. Washington, Printed by Gales & Seaton, 1823. 123 pp. HD.9724.A5

CENSUS OF 1830

Abstract

Census Office. 5th Census, 1830.

Abstract of the returns of the Fifth Census, showing the number of free people, the number of slaves, the federal or representative number; and the aggregate of each county of each State of the United States. Washington, Printed by D. Green, 1832. 51 pp. HA 201.1830.D2

General

Fifth Census; or, Enumeration of the inhabitants of the United States. 1830. To which is prefixed a schedule of the whole number of persons within the several districts of the United States, taken according to the acts of 1790, 1800, 1810, 1820. Washington, Printed by D. Green, 1832. 2 vol. Vol. I, 27 pp.; Vol. II, 163 pp. HA 201.1830B

Department of State.

Statistical view of the population of the United States from 1790 to 1830, inclusive. Furnished by the Department of State, in accordance with resolutions of the Senate of the United States on the 26th February, 1833, and 31st March, 1834. Washington, Printed by D. Green, 1835. iii, 216 pp. HA 195. U5. 1835

CENSUS OF 1840

Compendium

Census Office. 6th Census, 1840.

Compendium of the enumeration of the inhabitants and statistics of the United States, as obtained at the Department of State, from the returns of the Sixth Census, by counties and principal towns . . . to which is added an abstract of each preceding census. Prepared at the Department of State. Washington, Blair and Rives, 1841. 375 pp. HA 201.1840.C

General

Sixth Census, or enumeration of the inhabitants of the United States as corrected at the Department of State, in 1840. Published, by authority of an act of Congress, under the direction of the Secretary of State. Washington, Blair and Rives, 1841. 476 pp. HA 201.1840.B1.

Agriculture and Manufacturing

Statistics of the United States of America: 1840. Washington, Blair and Rives, 1841. 410 pp.
Agriculture, manufactures, mines and quarries, fisheries, forest products.

CENSUS OF 1850

Compendium

Census Office. 7th Census, 1850.
Statistical view of the United States, embracing its territory, population—white, free colored, and slave—moral and social condition, industry, property, and revenue; the detailed statistics of cities, towns and counties; being a compendium of the Seventh Census, to which are added the results of every previous census, beginning with 1790, in comparative tables, with explanatory and illustrative notes, based upon the schedules and other official sources of information . . . Washington, A. O. P. Nicholson, Public Printer, 1854. 400 pp. HA 201.1850.C

General

The Seventh Census of the United States: 1850. Embracing a statistical view of each of the States and territories, arranged by counties, towns, etc. . . . and an appendix embracing notes upon the tables of each of the States, etc. . . . Washington, R. Armstrong, Public Printer, 1853. cxxxvi, 1022 pp. HA 201.1850.B1

Manufacturing

Message of the President of the United States, communicating a digest of the statistics of manufactures according to the returns of the Seventh Census . . . [Washington, 1859] 143 pp. HA 201.1850.B3

Other

Mortality statistics of the Seventh Census of the United States, 1850 . . . with sundry comparative and illustrative tables. . . . Washington, A. O. P. Nicholson, Public Printer, 1855. 303 pp. HA 201.1850.B2

CENSUS OF 1860

General

Census Office. 8th Census, 1860.
Statistics of the United States (including mortality, property, &c.) in 1860; comp. from the original returns and being the final exhibit of the Eighth Census. Washington, Govt. Printing Office, 1866. lxvi, 584 pp. HA 201.1860.B4
HB 3505.A3

Population

Population of the United States in 1860; compiled from the original returns of the Eighth Census. Washington, Govt. Printing Office, 1864. cvii, 694 pp.
HA 201.1860.B1

Agriculture

Agriculture of the United States in 1860; compiled from the original returns of the Eighth Census . . . Washington, Govt. Printing Office, 1864. clxxii, 292 pp.
HA 201.1860.B3

Manufacturing

Manufactures of the United States in 1860; compiled from the original returns of the Eighth Census . . . Washington, Govt. Printing Office, 1865. ccxvii, 745 pp.
HA 201.1860.B2

CENSUS OF 1870

Compendium

Census Office. 9th Census, 1870.
A compendium of the Ninth Census (June 1, 1870), compiled pursuant to a concurrent resolution of Congress, and under the direction of the Secretary of the Interior, by Francis A. Walker, Superintendent of Census. Washington, Govt. Printing Office, 1872. vii, 942 pp. HA 201.1870.C

General

Census reports. Compiled from the original returns of the Ninth Census (June 1, 1870). 3 vol. Washington, Govt. Printing Office, 1872.
HA 201.1870.B1

Vol. I. The statistics of the population of the United States, embracing the tables of race, nationality, sex, selected ages, and occupations. To which are added the statistics of school attendance and illiteracy, of schools, libraries, newspapers, and periodicals, churches, pauperism and crime, and of areas, families and dwellings.

Vol. II. The vital statistics of the United States, embracing the tables of deaths, births, sex, and age, to which are added the statistics of the blind, the deaf and dumb, the insane, and the idiotic.

Vol. III. The statistics of wealth and industry of the United States, embracing the tables of wealth, taxation, and public indebtedness of agriculture; manufactures; mining; and the fisheries. With which are reproduced, from the volume on population, the major tables of occupations.

Population

Statistics of population. Tables I to VIII inclusive . . . Washington, Govt. Printing Office, 1872. 391 pp. HA 201.1870.B2

Contents: I. Population, 1870–1790, by States and territories, as white, free colored, slave, Chinese, and Indian. II. Population, 1870–1790, in each State, by counties, as white, free, colored, slave, Chinese, and Indian. III. Population, 1870–1850, in each State, by civil divisions less than counties, as white and colored. . . IV. Nativity (general) 1870–1850, foreign parentage, 1870, by States and territories. V. Nativity (general) 1870–1860, foreign parentage, 1860, in each State, by counties. VI. Nativity, (special) 1870, by States and territories. VII. Nativity (selected special) 1870, in each State, by counties. VIII. Nativity (special) 1870, fifty principal cities.

Statistical Atlas

Statistical atlas of the United 'States based on the results of the Ninth Census, 1870 . . . [New York], J. Bien, Lith., 1874. Separately paged. HA 201.1870.B2

CENSUS OF 1880

Compendium

Census Office. 10th Census, 1880.

Compendium of the Tenth Census (June 1, 1880), compiled pursuant to an act of Congress approved August 7, 1882 . . . Washington, Govt. Print. Office, 1883. lxxvi, 1769 pp. HA 201.1880c

General

Census reports, Tenth Census, June 1, 1880. Washington, Govt. Print. Office, 1883–88. 22 vol. HA 201.1880.B1

Vol. I. Statistics of the population . . . embracing extended tables of the population of states, counties, and minor civil divisions, with distinction of race, sex, age, nativity, and occupations: together with summary tables, derived from other census reports, relating to newspapers and periodicals; public schools and illiteracy; the dependent, defective, and delinquent classes, etc. (Introduction: General discussion of the movements of population. 1790–1880).

Vol. II. Manufactures: Remarks on the statistics of manufactures. General statistics. Tabular statements. Power used in manufactures. The factory system. Manufactures of interchangeable mechanism. Hardware, cutlery, and edge tools. Iron and steel production. Silk manufacture. Cotton manufacture. Woolen manufacture. Chemical products and salt. Glass manufacture.

Vol. III. Productions of agriculture: Remarks on the statistics of agriculture. General statistics: Tabular statements. Cereal production. Flour-milling. Tobacco culture. Manufacture and movement of tobacco. Meat production.

Vol. IV. Agencies of transportation: Statistics of railroads. Steam navigation. Canals. Telegraphs and telephones. Postal telegraphs.

Vol. V–VI. Cotton production in the United States: also embracing agricultural and physico-geographical descriptions of the several cotton states and of

California. (Pt. 1) General discussion of cotton production. Cotton production in the Mississippi Valley and Southwestern States; (Pt. 2) Cotton production in the Eastern Gulf, Atlantic, and Pacific States.

Vol. VII. Valuation, taxation, and public indebtedness.

Vol. VIII. The newspaper and periodical press. Alaska: its population, industries, and resources. The seal islands of Alaska. Shipbuilding industry in the United States.

Vol. IX. Forests of North America (exclusive of Mexico). Portfolio of 16 folded maps.

Vol. X. Production, technology, and uses of petroleum and its products. The manufacture of coke. Building stones of the United States, and statistics of the quarry industry for 1880.

Vol. XI–XII. Mortality and vital statistics. Portfolio of plates and diagrams.

Vol. XIII. Statistics and technology of the precious metals.

Vol. XIV. The United States mining laws and regulations thereunder, and State and territorial mining laws, to which are appended local mining rules and regulations.

Vol. XV. Mining industries of the United States (exclusive of the precious metals) with special investigations into the iron resources of the republic and into the cretaceous coals of the Northwest.

Vol. XVI–XVII. Statistics of power and machinery employed in manufactures. Water power of the United States.

Vol. XVIII–XIX. Social statistics of cities: (Pt. 1) The New England and the Middle States. (Pt. 2) The Southern and Western States.

Vol. XX. Statistics of wages in manufacturing industries; with supplementary reports on the average retail prices of necessaries of life, and on trade societies, and strikes and lockouts.

Vol. XXI. Defective, dependent, and delinquent classes of the population of the United States.

Vol. XXII. Power and machinery employed in manufactures.

CENSUS OF 1890

Abstract and Compendium

Bureau of the Census. 11th Census, 1890.

Abstract of the Eleventh Census: 1890 . . . Washington, Govt. Printing Office, 1894. vi, 250 pp. HA 201.1890.D1

Compendium of the Eleventh Census: 1890. Washington, Govt. Print. Office, 1892–97. 3 vol. HA201.1890.C

Vol. I. Population.

Vol. II. Vital and social statistics. Educational and church statistics. Wealth, debt, and taxation. Mineral industries. Insurance. Foreign-born population. Manufactures.

Vol. III. Population. Agriculture. Manufactures (totals for States and industries). Fisheries. Transportation. Wealth, debt, and taxation. Real estate mortgages. Farms and homes: Proprietorship and indebtedness. Indians (taxed and not taxed).

General

Census reports. Eleventh Census: 1890. Washington, Govt. Printing Office, 1892–97. 15 vol. HA 201.1890.B1

Published also as Pts. 1, 4, 7, 14–15, 17–18, 20–26, 28 of House Mis. Doc. 340, 52d Cong., 1st Sess. The Compendium (3 vol.) was published as Pt. 6, Report on education as Pt. 9, and the Statistical atlas as Pt. 29 of this document.

Vol. I. Population, 1895–97. 2 parts.

Vol. II. Insane, feebleminded, deaf and dumb, and blind . . . 1895.

Vol. III. Crime, pauperism, and benevolence . . . Pt. I. Analysis. 1896. Pt. II. General tables. 1895.

Vol. IV. Vital and social statistics . . . Pt. I. Analysis and rate tables. 1896. Pt. II. Vital statistics; cities of 100,000 population and upward. 1896. Pt. III–IV. Statistics of deaths. 1894–95.

Vol. V. Agriculture, etc.: Statistics of agriculture. 1895. Agriculture by irrigation in the western part of the United States . . . 1894. Statistics of fisheries. 1894.

Vol. VI. Manufacturing industries: Pt. I. Totals for States and industries. Pt. II. Statistics of cities. Pt. III. Selected industries. 1896.

Vol. VII. Mineral industries. 1892.

Vol. VIII. Population and resources of Alaska. 1893.
Vol. IX. Statistics of churches . . . 1894.
Vol. X. Indians taxed and Indians not taxed in the United States (except Alaska) 1894.
Vol. XI. Insurance business . . . Pt. I. Fire, marine, and inland insurance. Pt. II. Life Insurance. 1894–95.
Vol. XII. Real estate mortgages . . . 1895.
Vol. XIII. Farms and homes: Proprietorship and indebtedness . . . 1896.
Vol. XIV. Transportation business. Pt. I. Transportation by land. 1895. Pt. II. Transportation by water. 1894.
Vol. XV. Wealth, debt, and taxation. Pt. I. Public debt. Pt. II. Valuation. 1892–95.

Statistical Atlas

Statistical atlas of the United States, based upon the results of the Eleventh Census. Washington, Govt. Printing Office, 1898. 69 pp. HA 201.1890.E

Other

Report on the social statistics of cities in the United States at the Eleventh Census: 1890. Washington, Govt. Printing Office, 1895. vii, 137 pp. HA 205.A35.1890

CENSUS OF 1900[4]

Abstract

Census Office. 12th Census, 1900.
Abstract of the Twelfth Census of the United States, 1900. Washington, Govt. Printing Office, 1902. xiii, 395 pp. HA 201.1900D

General

Census reports . . . Twelfth Census of the United States, taken in the year 1900. Washington, Census Office, 1901–02. 10 vol. HA201.1900.B1
Vol. I–II. Population.
Vol. III–IV. Vital statistics. Pt. I. Analysis and ratio tables. Pt. II. Statistics of deaths.
Vol. V–VI. Agriculture. Pt. I. Farms, livestock and animal products. Pt. II. Crops and irrigation.
Vol. VII–X. Manufactures. Pt. I. United States by industries. Pt. II. States and territories. Pt. III–IV. Special reports on selected industries.

Bulletins

Bulletins 1–162. Washington, Govt. Printing Office, 1903. 162 Nos. (Nos. 1–4 issued by the United States Census Office.) HA 201.1900.A12
Bulletins of the Twelfth Census of the United States issued from October 6, 1900 to October 20, 1902. Washington, Census Office, 1900–02. HA 201.1900.A1
244 Nos. in 6 vols. Published at irregular intervals under title *Census bulletin.* Nos. 1–3, of an administrative character, were not issued for general distribution.

Occupations

Occupations at the Twelfth Census. Washington, Govt. Printing Office, 1904. cclxvi, 763 pp. HA 201.1900.B2
Statistics of women at work, based on unpublished information derived from the schedules of the Twelfth Census: 1900. Washington, Govt. Printing Office, 1907. 399 pp. HA 201.1900.B2

---

[4] The Twelfth Decennial Census was limited by law to the four fields of population, mortality, agriculture, and manufactures. Other subjects were to be covered after the work of the Decennial Census was completed. The Census Office was made a permanent agency of government in 1902. Hence many subjects previously covered in the Decennial Censuses became inter-censal inquiries, and special statistical reports replaced the former decennial publications. For a list of these publications, *see:* Bureau of the Census, Circular of information concerning census publications, 1790–1916. Jan. 1, 1917, Washington, Govt. Printing Office, 1917. 124 pp. Z 7554.U5U5.1916

Statistical Atlas

Statistical atlas. Washington, Census Office, 1903. 91 pp. HA 201.1900.B1

Other

Supplementary analysis and derivative tables. Twelfth Census of the United States. Washington, Govt. Printing Office, 1906. xvii, 1144 pp.
HA 201.1900.B2sc

Includes "Age," with bibliography, by Allyn A. Young; "A discussion of the vital statistics of the Twelfth Census," by John Shaw Billings; and "The Negro Farmer," by W. E. Burghardt DuBois.

CENSUS OF 1910

Abstract

Bureau of the Census.

Thirteenth Census of the United States taken in the year 1910. Abstract of the census. Statistics of population, agriculture, manufactures, and mining for the United States, the States, and principal cities. Washington, Govt. Printing Office, 1912. 569 pp. HA 201 1910.A2

General

Thirteenth Census of the United States taken in the year 1910. Reports, vol. I–XI . . . Washington, Govt. Printing Office, 1912–14. 11 vol.
HA 201.1910.A15

Vol. I–IV. Population: (I) General report and analysis. (II–III) Reports by States, with statistics for counties, cities and other civil divisions, Alabama to Wyoming, Alaska, Hawaii and Porto Rico. (IV) Occupational statistics.

Vol. V–VII. Agriculture, 1909 and 1910: (V) General report and analysis. (VI–VII) Reports by States, with statistics for counties, Alabama to Wyoming, Alaska, Hawaii and Porto Rico.

Vol. VIII–X. Manufactures, 1909: (VIII) General report and analysis. (IX) Reports by States, with statistics for principal cities. (X) Reports for principal industries.

Vol. XI. Mines and quarries, 1909: General report and analysis.

Life Tables

United States life tables, 1910. Washington, Govt. Printing Office. 1916. 65 pp. HG 8784.U6A5 1910

United States life tables, 1890, 1901, 1910, and 1901–1910. Explanatory text, mathematical theory, computations, graphs, and original statistics, also tables of United States life annuities, life tables of foreign countries, mortality tables of life insurance companies. Washington, Govt. Printing Office, 1921. 496 pp.
HG 8784.U6A5 1910a

Other

Indian population in the United States and Alaska. 1910. Washington, Govt. Printing Office, 1915. 285 pp. E. 77.U5

Statistics of the Indian population—Number, tribes, sex, age, fecundity, and vitality. Washington, Govt. Printing Office, 1913. 25 pp. E98.C3.U48 Folio

Negro population, 1790–1915. Washington, Govt. Printing Office, 1918. 855 pp. HA 205.A33

CENSUS OF 1920

Abstract

Bureau of the Census.

Abstract of the Fourteenth Census of the United States. 1920 . . . Washington, Govt. Printing Office, 1923. 1303 pp. HA 205.A5 1920g

General

Fourteenth Census of the United States taken in the year 1920 . . . Reports, Washington, Govt. Printing Office, 1921–23. 11 vol. HA 201 1920.A15

Vol. I–IV. Population. 1920. (I) Number and distribution of inhabitants. (II) General report and analytical tables. (III) Composition and characteristics of the population by States. (IV) Occupations.

Vol. V–VI. Agriculture. (V) General report and analytical tables. (VI) Reports for States, with statistics for counties and a summary for the United States and the North, South, and West. (Pt. 1) The northern States. (Pt. 2) The southern States. (Pt. 3) The western States and outlying possessions.
Vol. VII. Irrigation and drainage. General report and analytical tables, and reports for States, with statistics for counties.
Vol. VIII–X. Manufactures, 1919. (VIII) General report and analytical tables. (IX) Reports for States, with statistics for principal cities. (X) Reports for selected industries.
Vol. XI. Mines and quarries. 1919. General report and analytical tables and reports for States and selected industries.

Life Tables

United States abridged life tables, 1919–1920. Washington, Govt. Print. Office, 1923. 84 pp. HG 8784.U6A5 1920

Monographs

Census Monographs, I–XI. Washington, Govt. Printing Office, 1922–1931. HA 201 1920.A2

I. Increase of population in the United States, 1910–1920 . . . 1922. 255 pp. HA 205.R6
II. Mortgages on homes . . . 1923. 277 pp. HD 7293.A3A2 1920h
III. The integration of industrial operation . . . 1924. 272 pp. HD 9725.T5
IV. Farm tenancy in the United States . . . 1924. 247 pp. HD 1511.U5G6
V. School attendance in 1920 . . . 1924. xx, 285 pp. LC 143.R6
VI. Farm population of the United States . . . 1926. xi, 536 pp. HB 2385.T7
VII. Immigrants and their children, 1920 . . . 1927. xvi, 431 pp. JV 6455.C3
VIII. The growth of manufactures, 1899 to 1923 . . . 1928. 205 pp. HD 9724.D3
IX. Women in gainful occupations, 1870 to 1920 . . . 1929. xvi, 416 pp. HD 6093.H5
X. Earnings of factory workers, 1899 to 1927 . . . 1929. xvi, 424 pp. HD 4975.B7
XI. Ratio of children to women, 1920 . . . 1931. ix, 242 pp. HB 915.T5

Statistical Atlas

Statistical atlas of the United States. Washington, Govt. Printing Office, 1925, iii, 476 pp. HA 205.A4 1925

Other

The blind population of the United States, 1920 . . . A statistical analysis of the data obtained at the Fourteenth Decennial Census, Washington, Govt. Printing Office, 1928. 191 pp. HV 1876 1920a
The deaf-mute population of the United States, 1920. A statistical analysis of the data obtained at the Fourteenth Decennial Census . . . 1928. 153 pp. HV 2530.U6 1920
Children in gainful occupations at the Fourteenth Census of the United States . . . 1924. 276 pp. HD 6250.U3A3 1920a
Farm population of selected counties . . . 1924. 238 pp. HB 2385.A5
Mortality rates, 1910-1920, with population of the federal censuses of 1910 and 1920 and intercensal estimates of population. 1923. 681 pp. HA 201 1920.A6
Negroes in the United States, 1920–32. 1935. xvi, 845 pp. HA 205.A33 1920–32
This report supplements the volume "Negro population in the United States, 1790–1915," published in 1918.

CENSUS OF 1930

Abstract

Bureau of the Census.
Fifteenth Census of the United States: 1930. Abstract . . . Washington, Govt. Printing Office, 1933. viii, 968 pp. HA 201 1939.A3Z3
Forms part of the series of Reports of the Fifteenth Decennial Census.

General

Fifteenth Census of the United States: 1930. Reports. Washington, Govt. Printing Office, 1931–34. 32 vol. HA 201 1930.A3

Agriculture. (I) Farm acreage and farm values by townships or other minor civil divisions. (II) Reports by States, with statistics for counties and a summary for the United States. (Pt. 1) The northern States. (Pt. 2) The southern States. (Pt. 3) The western States. (III) Type of farm. Reports by States, with statistics for counties and a summary for the United States. (Pt. 1) The northern States. (Pt. 2) The southern States. (Pt. 3) The western States. (IV) General report, statistics by subjects. HA 201.1930.A3A5

Construction industry. Reports by States, with statistics for counties and cities of 100,000 population and over, a summary for the United States, and a study of the location and agencies of the construction industry. HA 201.1930.A3C6

Distribution. (I) Retail distribution. (Pt. 1) Summary for the United States, and statistics for counties and incorporated places of 1,000 population and over. (Pt. 2) Reports by States. Alabama-New Hampshire. (Pt. 3) Reports by States. New Jersey-Wyoming. (II) Wholesale distribution. State reports with statistics for cities and a summary for the United States including county statistics. HA 201.1930.A3D5

Drainage of agricultural lands. Reports by States with statistics for counties, a summary for the United States, and a synopsis of drainage laws. HA 201.1930.A3D7

Horticulture. Statistics for the United States and for States. 1929 and 1930. HA 201.1930.A3H6

Irrigation of agricultural lands. General reports and analytical tables. Reports by States with statistics for counties, and a summary for the United States. HA 201.1930.A317

Manufactures: 1929. (I) General report. Statistics by subjects. (I) Reports by industries. (III) Reports by States. Statistics for industrial areas, counties, and cities. HA 201.1930.A3M3

Metropolitan districts. Population and area. HA 201.1930.A3M4

Mines and quarries: 1929. General report and reports for States and for industries. HA 201.1930.A3M5

Outlying territories and possessions. Number and distribution of inhabitants, composition and characteristics of the population, occupations, unemployment and agriculture. HA 201.1930.A307

Population. (*I*) *Number and distribution of inhabitants.* Total population for States, counties, and townships or other minor civil divisions; for urban and rural areas; and for cities and other incorporated places. (*II*) *General report. Statistics by subjects.* (*III*) *Reports by States, showing the composition and characteristics of the population* for counties, cities, and townships or other minor civil divisions. (Pt. 1) Alabama-Missouri. (Pt. 2) Montana-Wyoming. (*IV*) *Occupations by States.* Reports by States, giving statistics for cities of 25,000 or more. (*V*) *General report on occupations.* (*VI*) *Families.* Reports by States, giving statistics for families, dwellings and homes, by counties, for urban and rural areas and for urban places of 2,500 or more. HA 201.1930.A3P6

Unemployment. (*I*) *Unemployment returns by classes* for States and counties, for urban and rural areas, and for cities with a population of 10,000 or more. (*II*) *General report.* Unemployment by occupation, April, 1930, with returns from the special census of unemployment, January, 1931. HA 201.1930.A3U6

Life Tables

United States life tables. 1929 to 1931, 1920 to 1929, 1919 to 1921, 1909 to 1911, 1901 to 1910, 1900 to 1902 . . . Washington, Govt. Printing Office, 1936. vi, 57 pp. HG 8784.U6A5 1930

Other

Age of the foreign-born white population by country of birth. Washington, Govt. Printing Office, 1933. vi, 77 pp. HB 3015.A5 1930c

The Indian population of the United States and Alaska . . . 1937. vi, 238 pp. E 77.U5 1937

A social-economic grouping of gainful workers of the United States. Gainful workers of 1930 in social-economic groups, by color, nativity, age, and sex, and by industry, with comparative statistics for 1920 and 1910. With appendix. 1938. 279 pp. HD 8064.A5 1930a

Special report on foreign-born white families by country of birth of head. With an appendix giving statistics for Mexican, Indian, Chinese, and Japanese families . . . 1933. 217 pp. HB 3015.A5 1930b

CENSUS OF 1940

POPULATION

Bureau of the Census.

Sixteenth Census of the United States: 1940. Reports. Washington, Govt. Printing Office, 1940–1943.

Population. (I) Number of inhabitants. Total population for States, counties, and minor civil divisions; for urban and rural areas; for incorporated places; for metropolitan districts; and for census tracts. *(II) Characteristics of the population.* Sex, age, race, nativity, cirizenship, country of birth of foreign-born white, school attendance, education, employment status, class of worker, major occupation group, and industry group. Reports by States. 7 pts. *(III) The labor force.* Occupation, industry, employment, and income. Reports by States. 5 pts. *(IV) Characteristics by age.* Marital status, relationship, education, and citizenship. Reports by States. 4 pts.

The statistics presented in each of the above volumes were first published, and are still available, in the form of a series of separate bulletins for each State and a summary bulletin for the United States. In their separate form, the sections of Vol. I comprise the first series of Population bulletins for the States; those of Vol. II comprise the second series . . . etc.

Population. Internal migration, 1935 to 1940. A series of reports based on tabulations from the 1940 population census.

Color and sex of migrants. Washington, Govt. Printing Office, 1943. viii, 490 pp. (In press)

Other reports, including one on State of birth by State of residence, are planned in this series.

Population. Characteristics of the nonwhite population by race. Washington, Govt. Printing Office, 1943. vi, 112 pp.

Population. Special report on the institutional population 14 years old and over. Characteristics of inmates in penal institutions, and institutions for the delinquent, defective, and dependent. Washington, Govt. Printing Office, 1943. iv, 361 pp.

Population. Nativity and parentage of the white population. A series of three bulletins based on a 5-percent sample of the population. Washington, Govt. Printing Office, 1943. (In press)

General characteristics. Age, marital status and education for States and large cities. vi, 279 pp.

Country of origin of the foreign stock. By nativity, citizenship, age, and value or rent of home, for States and large cities. vi, 122 pp.

Mother tongue. By nativity, parentage, country of origin, and age, for States and large cities. vi, 58 pp.

Population. Differential fertility: 1940 and 1910. A series of reports based on sample tabulations of the population data for 1940 and 1910.

Fertility for States and large cities. Age, color, and marital status of women 15 to 74 years old, classified by number of children ever born, and number of children under 5 years old and 5 to 9 years old. Women by age at marriage and duration of marriage . . . Washington, Govt. Printing Office, 1943. viii, 281 pp. (In press)

Other reports are planned in this series.

Population. Education, occupation, and household relationship of males 18 to 44 years old. Prepared by the Division of Population, Bureau of the Census, of the Department of Commerce, in cooperation with the Special Services Division of the War Department. Washington, Govt. Printing Office, 1943, vi, 59 pp.

Population. The labor force (sample statistics). A series of six reports based on tabulations of sample statistics from the 1940 population census. Washington, Govt. Printing Office, 1943. (In press)

Employment and personal characteristics. vi, 177 pp.
Employment and family characteristics of women. vi, 212 pp.
Wage or salary income in 1939. vi, 194 pp.
Occupational characteristics. vi, 256 pp.
Industrial characteristics. iv, 174 pp.
Usual occupation. iv, 63 pp.

Population. Characteristics of persons not in the labor force, 14 years old and over. Age, sex, color, household relationship, months worked in 1939, and usual major occupation group. Washington, Govt. Printing Office, 1943. vi, 117 pp.

Population. Comparative occupation statistics for the United States, 1870 to 1940. A comparison of the 1930 and 1940 census occupation and industry classification and statistics; a comparable series of occupation statistics, 1870 to 1930; and a social-economic grouping of the labor force, 1910 to 1940. Washington, Govt. Printing Office, 1943. (In press)

POPULATION AND HOUSING

Population and Housing. Statistics for census tracts. Washington, Govt. Printing Office, 1941–1943.

A series of bulletins, one for each of 60 tracted cities, presenting basic population and housing statistics by census tracts. Accompanying maps show the principal boundary streets for each tract. Population items include sex, age, race, nativity, citizenship, country of birth, education, employment status, class of worker, and occupation. Housing items include occupancy, tenure, value or monthly rent, type of structure, state of repair and plumbing equipment, size of household, race of head of household, persons per room, radio, refrigeration equipment, and heating fuel by type of heating equipment.

Population and Housing. Families. A series of reports based on tabulations of sample statistics from the 1940 population and housing censuses. Washington, Govt. Printing Office, 1943. (In press)

General characteristics. States, cities of 100,000 or more, and metropolitan districts of 200,000 or more. Tenure, size of family, number of children, labor-force status of children, number of lodgers and subfamilies, number of persons in labor force, family employment status, family wage or salary income in 1939 . . . vi, 332 pp.

Types of families. Regions and cities of 1,000,000 or more. vi, 221 pp.

Employment status. Regions and cities of 1,000,000 or more. v, 110 pp.

Family wage or salary income in 1939. Regions and cities of 1,000,000 or more. iv, 156 pp.

Tenure and rent. Regions, cities of 1,000,000 or more, and metropolitan districts of 500,000 or more. vi, 141 pp.

Income and rent. Regions, and metropolitan districts of 1,000,000 or more. vi, 237 pp.

Characteristics of rural-farm families. Regions and divisions. Tenure, occupation of head, and value or rent, cross-classified by selected housing characteristics, family characteristics, and characteristics of head. iv, 82 pp.

HOUSING

Housing. (*I*) *Data for small areas.* Selected housing statistics for States, counties, and minor civil divisions; for urban and rural areas; for incorporated places; and for metropolitan districts. Reports by States, 2 pts. (*II*) *General characteristics.* Occupancy and tenure status, value of home or monthly rent, size of household and race of head, type of structure, exterior material, year built, conversion, state of repair, number of rooms, housing facilities and equipment, and mortgage status. Reports by States. 5 pts. (*III*) *Characteristics by monthly rent or value.* These statistics form the basis for determining the relationship between rent or value and such characteristics as type and age of structure, state of repair, number of rooms, size of household, race of head, persons per room, housing facilities and equipment, and mortgage status. Reports by States. 3 pts. (*IV*) *Mortgages on owner-occupied nonfarm homes.* First mortgages by amount outstanding, type of payment, frequency and amount of payment, interest rate, and holder of mortgage. All mortgaged properties by value of property, estimated rental value, year built, and color of occupants . . . Supplements A and B show mortgage data for homes built in 1935–1940, and homes occupied by nonwhite owners, respectively. Reports by States. 3 pts.

The statistics presented in each of the above volumes were first published, and are still available, in the form of a series of separate bulletins for each State and a summary bulletin for the United States. In their separate form, the sections of Vol. I comprise the first series of Housing bulletins for the States; those of Vol. II comprise the second series . . . etc.

Housing. Block statistics. Supplements to the first series of Housing bulletins for States (Housing, Vol. I). Washington, Govt. Printing Office, 1941–1943.

A series of supplemental bulletins, one for each of the 191 cities which in 1930 had a population of 50,000 or more, showing selected housing statistics for the city by blocks. Data are also presented by census tracts for certain cities, and by wards for other cities. Each bulletin includes a map of the given city by blocks.

Housing. Analytical maps. Block statistics. Prepared by the New York City WPA War Services and the Bureau of the Census; distributed by the Bureau of the Census, 1942 to date.

A series of bulletins, one for each city of 100,000 inhabitants or more, comprising sets of analytical maps presenting in graphic form various housing characteristics in different areas of the city. Includes maps for average rent, major repairs or bathing equipment, year built, nonwhite households, persons per room, owner occupancy, and mortgage status.

Agriculture

Agriculture. (I) First and second series State reports. Statistics for counties. Farms and farm property, with related information for farms and farm operators, livestock and livestock products, and crops. 6 pts. *(II) Third series State reports.* Statistics for counties. Value of farm products, farms classified by major sources of income, farms classified by value of products. 3 pts. *(III) General report.* Statistics by subjects. Washington, Govt. Printing Office, 1942–1943. HA 201.1940.A4

Agriculture. Cows milked and dairy products. Washington, Govt. Printing Office, 1943. vi, 556 pp. HD 9275.U6A45 1940A

Agriculture. Cross-line acerage. Washington, Govt. Printing Office, 1943. iv, 311 pp.

Agriculture. Drainage of agricultural lands. Washington, Govt. Printing Office, 1942. xvi, 683 pp. HD1683.U4A5. 1940

Agriculture. Irrigation of agricultural lands. Washington, Govt. Printing Office, 1942. lxv, 689 pp. HD1725.A5. 1940

Agriculture. Special cotton report. Washington, Govt. Printing Office, 1943. xv, 265 pp.

Agriculture. Farm characteristics by value of products. An analysis of specified farm characteristics for farms classified by total value of products. Washington, Govt. Printing Office, 1943. (In press)

A report based on a 2-percent sample of farms reported in the 1940 census of agriculture and prepared cooperatively by the Bureau of Agricultural Economics and the Bureau of the Census.

Business

Census of business, 1939. (I) Retail trade. (Pt. 1) Not yet issued. (Pt. 2) Commodity sales and analysis by sales size . (Pt. 3) Kinds of business, by area, States, counties, and cities. (II) Wholesale trade. (III) Service establishments, places of amusement, hotels, tourist courts and tourist camps. (IV) Construction. (V) Distribution of manufacturers' sales. Washington, Govt. Printing Office, 1942–1943.

Manufactures

Manufactures, 1939. (I) Statistics by subjects. (II) Reports by industries. 2 pts. (III). Reports for States and outlying areas. Washington, Govt. Printing Office, 1942. HD 9724.A4. 1940b

Life Tables

United States life tables, 1930–1939. Preliminary. For white and nonwhite by sex. Washington, 1942. 14 pp.

These are the first life tables based on the results of the 1940 census of population and on the mortality experience of the decade. Life table values are shown for every year of age.

United States abridged life tables, 1930–1939. Preliminary. By geographic divisions, color and sex. Washington, 1942. 15 pp.

These life tables, the first separate tables for geographic divisions covering the entire United States, are based on the mortality of 1930–1939. Life table values are also shown for the death registration States of 1920 and 1930.

United States abridged life tables, 1939, urban and rural, by regions, color, and sex. Washington, 1943. 18 pp.

Abridged life tables are presented by urban and rural residence, color, and sex, for the continental United States and three major regions. Life table values are shown separately for the following urban-rural classes: (1) cities of 100,000 or more, (2) other urban places, and (3) rural territory.

Other

Areas of the United States: 1940. Washington, Govt. Printing Office, 1942. 465 pp.

Unincorporated Communities. United States, by States. Washington, Govt. Printing Office, 1943. iv, 32 pp.

Total population of unincorporated communities having 500 or more inhabitants for which separate figures could be compiled.

CURRENT NATIONAL VITAL STATISTICS

General

Bureau of the Census.

Vital statistics of the United States, 1940. (Part I) Natality and mortality data for the United States tabulated by place of occurrence, with supplemental tables for Hawaii, Puerto Rico, and the Virgin Islands. (Part II) Natality and mortality data for the United States tabulated by place of residence. 1941. Part I, iv, 657 pp. Part II, iii, 283 pp. HA 203.A22.

The summary and rate tables in Part I are as follows: (I) Population, number of births and deaths, and crude rates: the registration States, United States, 1900–1940. (II) Enumerated population by cities and rural areas, and by race: United States and each State, April 1, 1940. (III) Enumerated population by age, race, and sex: United States, April 1, 1940. (IV) Population: United States and each State, 1930–1940. (V) Population, number of births and deaths, and crude rates: specified countries, 1938. (VI) Population, number of births and deaths, and crude rates: specified countries and selected years. (VII) Crude birth rates . . .: United States, each division and State, 1916–1940. (VIII) Crude stillbirth ratios . . .: United States, each division and State, 1922–1940. (IX) Crude death rates . . .: United States, each division and State, 1916–1940. (X) Infant mortality rates . . .: United States, each division and State, 1916–1940. (XI) Deaths from each cause and crude death rates: United States, 1939 and 1940. (XII) Crude death rates . . . for selected causes of death: United States, each division and State, 1940. (XIII) Population, births, deaths, infant mortality, and stillbirths: United States, each State by urban and rural areas and by race, and individual counties and cities by race, 1940. There are also general tables on live births, stillbirths, total deaths, and infant deaths, and a special joint-cause-of-death table. Selected tables are included for Hawaii, Puerto Rico, and the Virgin Islands.

Part II gives tabulations by place of residence rather than place of occurrence. Fertility data in this part are classified by geographic area; race, nativity, and age of parents; sex; number of children; and type of attendance. Mortality data are classified by geographic area, cause of death, sex, race, and institution.

Manual of the International List of Causes of Death (Fifth Revision) and Joint Causes of Death (Fourth Edition) 1939. 452 pp. 1940.

The International List, as adopted for use in the United States, is based on the Fifth Decennial Revision by the International Commission, Paris, October 3–7, 1938. The Manual of Joint Causes of Death is incorporated in this volume.

Physicians' Handbook on Birth and Death Registration, 1939. iii, 94 pp.

Vital Statistics—Special Reports

This series consists of eighteen volumes, to date. Even-numbered volumes, through Vol. 10, are annual vital statistics summaries for each State. Odd-numbered volumes, through Vol. 9, are selected reports and studies on special topics. Vol. 11 was a special series of State reports on motor-vehicle accident mortality. Vols. 12 and 15 were continuations of odd-numbered Volumes 1 through 9. Vol. 14 was a continuation of the State summaries. Vol. 13 was a special volume devoted to detailed reports on hospital and institutional facilities.

Current volumes are as follows:

Vol. 16. Mortality summary for registration States. Washington, July 7, 1942, to May 15, 1943.

The first report consists of an introduction and basic population data. Other reports present statistics on mortality as follows: 2. All causes. 3. Typhoid and paratyphoid fever. 4. Scarlet fever. 5. Whooping cough. 6. Diphtheria. 7. Tuberculosis (all forms). 8. Tuberculosis of the respiratory system. 9. Dysentery. 10. Malaria. 11. Syphilis (all forms). 12. Measles. 13. Cancer (all forms). 14. Cancer of female genital organs. 15. Cancer of the breast. 16. Diabetes mellitus. 17. Exophthalmic goitre. 18. Pellagra (except alcoholic). 19. Alcoholism (ethylism). 20. Intracranial lesions of vascular origin. 21. Diseases of the heart (all forms). 22. Chronic rheumatic diseases of the heart. 23. Diseases of the coronary arteries and angina pectoris. 24. Diseases of the heart. 25. Bronchitis. 26. Pneumonia (all forms), and influenza. 27. Bronchopneumonia. 28. Lobar pneumonia. 29. Pneumonia (unspecified). 30. Influenza. 31. Ulcer of stomach or duodenum. 32. Diarrhea, enteritis, etc. 33. Appendicitis. 34. Hernia and intestinal obstruction. 35. Cirrhosis of the liver. 36. Biliary calculi and other diseases of the gallbladder and biliary ducts. 37. Nephritis (all forms). 38. Senility. 39. Suicide. 40. Homicide. 41. All accidents. 42. Motor-vehicle accidents. 43. Ill-defined and unknown causes. 44. Locomotor ataxia (tabes dorsalis). 45. General paralysis of the insane. 46. Aneurysm of the aorta. 47. Syphilis (other forms). 48. Cerebrospinal (meningococcus) meningitis. 49. Smallpox. 50. Poliomyelitis and polioencephalitis (acute). 51. Diseases of the prostate. 52. Diseases of pregnancy, childbirth, and the puerperium. 53. Puerperal septicemia. 54. Puerperal toxemias. 55. Congenital malformations and diseases peculiar to the first year of life. 56. Congenital malformations. 57. Congenital debility. 58. Premature birth. 59. Injury at birth. 60. The infant (under 1 year of age). 61. The preschool child (1–4 years of age). 62. The school child (5–14 years of age). 63. The youth (15–24 years of age). 64. The young adult (25–44 years of age). 65. The middle-age adult (45–64 years of age). 66. The older ages (65 years and over). (Nos. 63–66 are in preparation.)

Vol. 17. Selected studies. Nov. 28, 1942 to May 15, 1943.

1. Announcement of Volume 17—selected studies.
2. A summary of natality and mortality data: United States, 1941.
3. Deaths from selected causes: United States, 1941.
4. Live births by person in attendance: United States, 1941.
5. Deaths from puerperal causes: United States, 1941.
6. Deaths from each cause: United States, 1939–1941.
7. Births, deaths, infant deaths, and stillbirths, each State: 1941.
8. Accident fatalities in the United States, 1941.
9. Marriage statistics—Resident brides and grooms by age: collection area, United States, 1940.
10. Live births by month: United States, 1941.
11. Stillbirths by person in attendance: United States, 1941.
12. Illegitimate live births by race: United States, 1938–1941.
13. Marriage statistics—Resident brides and grooms by previous marital status: collection area, United States, 1940.
14. Marriage statistics—Marriages occurring in collection area, by place of residence of brides and grooms, 1940.
15. Number of deaths and death rates by age, race, and sex, each State: 1941 (by place of residence).
16. Deaths in institutions by type of service and control, each State: 1941.
17. Deaths under 1 year from selected causes, by age and race, and rates by race: 1941 (by place of residence).
18. Studies in the completeness of birth registration—Part I. Completeness of birth registration: United States, December 1, 1939 to March 31, 1940.
19. Deaths by race, sex, and cause: United States, 1941.
20. Births, deaths, and infant deaths, each State, county, and city: 1941 (Births and deaths by place of residence; infant deaths by place of occurrence).
21. Births and deaths by place of occurrence and by place of residence, each State: 1941.
22. Marriage statistics—Marriages by racial type and by age of resident groom, by age of bride: collection area, 1940.
23. Marriage statistics—Resident brides by age and race: collection area, 1940.
24. Deaths and death rates for selected causes, by age, race, and sex: United States, 1941.

Vol. 18. State summaries of vital statistics, 1941. Jan. 5, 1943 to May 15, 1943. All tabulations except those for counties are by place of occurrence. Rates are based on revised population estimates for the years 1900 to 1939. Rates for 1940 and 1941 are based on the enumerated census populations for each State and on estimated populations for the District of Columbia and United States totals. The United States summary and the reports for each State have been issued.

OTHER

Weekly mortality index. Deaths in major cities for week ended . . . Vol. 14, 1943.

Data are presented for ninety cities, based on weekly telegraphic reports on death certificates received in each reporting city bureau.

Monthly vital statistics bulletin. Vol. 6, 1943.

Provisional birth, death, and infant death figures are given for each of forty States, and for the cities of Chicago, New Orleans, Baltimore, Boston, and New York, and for the Territory of Hawaii. The annual summary for 1942, contained in Vol. 5, No. 13, was issued March 30, 1943.

Current mortality analysis, based on the returns of a 10-percent sample of death certificates received in vital statistics offices. Vol. 1, 1943.

A discussion of mortality trends for the month is followed by tables on numbers of deaths and death ratios for selected causes, by geographic divisions. Variation charts are given for selected diseases and areas. No. 3, for January 1943, was available in the latter part of March.

Motor-vehicle accident deaths.

Vol. 7 was the last volume issued prior to the suspension of the series for the duration of the war, as announced on January 18, 1943. Statistics were given for geographic divisions, and were issued in weekly releases and quarterly summaries.

The Registrar. Washington, Bureau of the Census. Vol. 8, No. 5, May 1943.

This monthly publication contains news notes and reports from the Division of Vital Statistics.

Summary of motor-vehicle accident fatalities (quarterly) contains a compilation of statistics on motor-vehicle accident deaths based on a special transcript form filled out by State bureaus of vital statistics and State traffic authorities. An annual summary is issued about 6 months after the end of each year.

SPECIAL PUBLICATIONS

Vital statistics rates in the United States, 1900–1940, (in preparation) approximately 1000 pp. A detailed compilation of death rates by age, race, sex, cause, and geographic area for the period 1900–1940 and of birth rates by age of mother, race, number of children, and geographic area for the period, 1915–1940. Corresponding population tables are included.

Vital statistics of the United States, Part III, Supplement, 1939–1940, (in preparation) approximately 550 pp. A compilation of birth and death data for local areas tabulated by place of residence. Most of the figures shown are two-year totals. Six summary tables giving two-year average birth and death rates are included.

URUGUAY

Historical

The only population statistics available for Uruguay prior to the middle of the nineteenth century consist of the estimates of travelers, plus the results of municipal enumerations in Montevideo in 1813 and 1816, and an official estimate in 1829 for the country as a whole.[1] The first census to receive official recognition, that of 1852–53, was never published in any detail, although it appears to have been used as a basis for estimates. Some data from it were included for comparative purposes in the report on the Census of 1860.[2] A brief report on population by nationality and sex appears to have been all that was published from the Census of 1900. The last national census was taken in 1908. An industrial census was taken in 1937, but the brief reports on number of workers in industrial establishments are an insufficient basis for population estimates.

NATIONAL POPULATION CENSUSES

CENSUS OF 1852

No separate publications were located. See pp. 33–34 of the 1860 census for provincial populations from this census.

CENSUS OF 1860

Secretaría de hacienda.

"Censo de la población de la República Oriental del Uruguay mandado levantar en agosto de 1860." *Registro estadístico de la República Oriental del Uruguay, 1860.* Tomo primero, Segunda sección, pp. 7–75. Montevideo, Imprenta de la República, 1863. 220 pp. HA 1071

CENSUS OF 1900

Comisión del censo.

Primer resumen del censo levantado el 1° de marzo de 1900 en los departamentos de campaña. Nacionalidad y sexo de las personas censadas...Montevideo, Tip. Escuela nacional de artes y oficios, 1900. 19 pp. HA 1071.A3 1900

CENSUS OF 1908

Dirección general de estadística.

"Censo general de la República en 1908." *Anuario estadístico de la República Oriental del Uruguay, año 1908.* Tomo II, Parte III, Montevideo, Imprenta Artística y encuadernación de Juan J. Dornaleche, 1911. pp. 755 to 1260. HA 1071 1908

Partial list of data included: Population by nationality and sex for provinces and departments. Place of birth. Women according to number of children, marital status, and nationality. Population of departments by single years of age by nationality. Education by age, nationality and sex. Literacy by nationality and sex. Religion, vaccination, physical and mental deficiencies, occupations.

---

[1] For a résumé of the census history of Uruguay, *see*: Fonticelli, Eduardo J. "Las actividades estadísticas del Uruguay." pp. 623–641 in: Inter American Statistical Institute. *Op. cit.* Early estimates are summarized in the following work: Uruguay, Conseil national de statistique. La République Orientale de l'Uruguay, quelques renseignements statistiques, souvenir du Conseil national de statistique aux membres de la mission spéciale du gouvernement Français, presidée par M. Pierre Baudin. Montevideo, 1915. 66 pp., fold. tables, maps. HA 1075.A4. 1914

[2] Secretaría de hacienda. *Registro estadístico de la República Oriental de Uruguay, 1860.* Tomo primero. Montevideo, Imprenta de la República, 1863. 220 pp. HA 1071

In addition to the population census, there were censuses of agriculture, industry, housing and education.

OTHER NATIONAL CENSUSES

CENSUS OF INDUSTRY, 1937

Ministerio de industrias y trabajo. Dirección de estadística económica.
Censo industrial de 1936 realizado en el año 1937. Cuadro general. Montevideo, 1939. 7 pp. HC 231.A3
Number of establishments, workers, capital, wages, and productivity by provinces and industrial groups.

CURRENT NATIONAL VITAL STATISTICS

(Including Population Estimates)

Dirección general de estadística.
Anuario estadístico de la República Oriental del Uruguay, año 1939. Volumen 1, Tomo 46, Publicación 157. Montevideo, Imprenta nacional, 1941. 417, xcvii pp. Govt. Publ. R. R.
The section on *Población* gives estimated populations for departments as of Dec. 31, 1938, natural and migratory increase, 1939, and estimated population and density as of Dec. 31, 1939. Vital statistics from the *Registro del estado civil* are presented for departments, by years, 1930–1939 inclusive. The more detailed statistics for 1939 include deaths by cause for departments. The section on *Migración* summarizes international migration statistics for 1939. There is a section on *Higiene y salud pública.*
Síntesis estadística . . . año 1940. Publicación 156, No. 18. Montevideo, 1941. 215 pp., xl. Govt. Publ. R. R.
Part II, "Población," gives population estimates as of Dec. 31, 1939, based on the 1938 estimates plus or minus natural increase and migration. The increase of population is surveyed from 1889 to 1939, while vital statistics trends are presented for 1930 to 1939. Detailed statistics are presented for marriage, divorce, fertility and mortality, 1939. Part IV covers *Higiene y salud pública.*
The summary of the report of the *Dirección general de estadística* includes a descriptive article on selected aspects of mortality, and tables on estimated population and vital rates for 1940. There are tables on deaths, 1889, 1909, 1919, 1929, and 1939, by age; mortality for departments by sex, 1936–1940; mortality by marital status and nationality; and mortality by cause.

Dirección general del registro del estado civil.
El movimiento del estado civil y la mortalidad de la República Oriental del Uruguay en el año 1940. Anuario de la Dirección general del registro del estado civil. Montevideo, Imprenta nacional, 1941. 74 pp. HA 1073.A45.
Fertility data include births for departments by sex, and paternal nationality; marriage statistics for departments include classifications by age, education, and literacy. Detailed tabulations on general and infant mortality are also included.

CURRENT CITY VITAL STATISTICS

MONTEVIDEO

Montevideo, Municipalidad de.
Boletín mensual de estadística 39(453):1–7. May, 1941. HA 1089.M8A2
Estimated population, migration, vital statistics, for the city as a whole and for sections.
Dirección de censo y estadística, *Resumen anual de estadística municipal.* Vol. 22, 1924. Montevideo, Imprenta nacional, 1926. 281 pp. HA 1089.M8A3
In addition to *vital* and migration statistics for 1924, there are estimates of population, 1889–1924, and migration 1879–1924.

VENEZUELA

Historical

The first reported census of Venezuela is that taken by Bishop Martí between 1772 and 1784, covering about half of the area of the country.[1] A summary of the many estimates of counts of the late colonial and early national period was included in the three-volume *Memoria* published by the *Dirección general de estadística* in 1873. A discussion of validity and method is included.[2]

National population censuses were taken in 1873, 1881, 1891, 1920, 1926, 1936, and 1941. The value of the two earliest censuses has been questioned seriously because of inconsistencies between the two. Regulations were published for a census in 1910, but apparently it was never taken. The Census of 1936 was carried out by a General Commissariat of the Census created especially for the purpose. A nation-wide organization was developed, and census schedules were distributed 15 days before the census date. These schedules were filled in by the heads of households and collected on the day set for "taking" the census. Only preliminary data are available from the Census of 1941.

Vital statistics are now collected and tabulated by the section on Demography of the General Bureau of Statistics. The new system, instituted on January 1, 1937, provided that local registration records be sent directly to the General Bureau of Statistics.[3] It is hoped that this national system will permit the development of more accurate vital statistics than those formerly secured by the combination of the reports of the statistical services of the individual states.

NATIONAL POPULATION CENSUSES

EARLY COUNTS

Dirección general de estadística.

Memoria de la dirección general de estadística al Presidente de los Estados Unidos de Venezuela en 1873. Maracaibo, Librería, Picón, 1873. Vol. I, xvi, 311 pp. Vol. II, 320 pp. Vol. III, 288, xxxvii pp. HA 1091.1873

Vol. I presents estimates of the population of provinces and cities during the 18th century. Vol. II contains a chapter on population, pp. 245–264, which gives summary figures ascribed to official censuses of 1825, 1838, 1844, 1846, 1847, 1854, and 1857. Vol. III gives data from a census of Caracas taken in 1869.

---

[1] For a historical résumé of Venezuelan statistics, see: Vandellós, José A. "Las actividades estadísticas de Venezuela." pp. 651-676 in: InterAmerican Statistical Institute. *Op. cit.* See also: Ibid. "Ensayo de demografía Venezolana." Caracas, Lit. y Tip. Casa de especialidades, 1938. 48 pp. HB 3579.V3.

[2] Dirección general de estadística. *Memoria de la Dirección general de estadística al Presidente de los Estados Unidos de Venezuela en 1873.* Maracaibo, Libería, Picón, 1873. Vol. I, XVI, 311 pp. Vol. II, 320 pp. Vol. III, 288, xxvii pp. HA 1091.1873.

Vol. I discusses colonial censuses briefly and presents data for various provinces from censuses taken in 1822, 1825, 1829, and 1869. Census data are given for the cantons of the Province of Caracas for 1825, 1829, and 1833. Vol. II, pp. 245–264, surveys early estimates and counts. The appendix in Vol. III gives summary figures from an 1869 census for Caracas.

[3] Vandellós, José A. *Op. cit.*

CENSUS OF 1873

Junta directiva del censo.

Primer censo de la República de Venezuela. . . . Verificado en los días 7, 8 y 9 de noviembre de 1873. Primera parte. Caracas, Imprenta nacional, 1874. xxix, 584 pp. HA 1091 1873A

An introduction on the legal basis, decrees, etc., for the census is followed by provincial reports. The volume ends, "Fin de la primera parte," but no second part was located.

CENSUS OF 1881

Junta directiva del censo.

Segundo censo de la República de Venezuela. . . . Verificado en los días 27, 28 y 29 de abril de 1881. Caracas, Imprenta Bolívar, 1881. 34, 403, xxxii pp. HA 1091 1881

Primera parte: Decreto ordenando la formación del segundo censo. Segunda parte: [Tables.] Documentos.

CENSUS OF 1891

Junta directiva del censo.

Tercer censo de la República. Caracas, Imprenta y litografía del gobierno nacional, 1891. 4 vols. HA 1091.A7 1890

Tomo III has imprint: Caracas, Casa editorial de la opinión nacional, 1891. Verificado en los días 15, 16 y 17 de enero de 1891.

PLANS FOR CENSUS OF 1910 [4]

Laws, statutes, etc.

Censo de Venezuela 1910. Decreto ejecutivo. Caracas? Imprenta nacional, 1910. 30 pp. HA 1091.A7 1910

CENSUS OF 1920

The census of 1920 does not appear to have been published separately. The *Anuario estadístico de Venezuela, 1938,* gives summary data from it in a table entitled: "Cuadro comparativo de la población de los estados y principales municipios según los censos efectuados en la República." HA 1091.A4.1938

CENSUS OF 1926

Dirección general de estadística.

Quinto censo nacional de los Estados Unidos de Venezuela decretado el 15 de agosto de 1925, . . . y levantado en los días 31 de enero y 1°, 2 y 3 de febrero de 1926. Caracas, Tipografía universal, 1926. 4 vol. HA 1091.A7 1926

I. Anzoátegui, Apure, Aragua, Bolívar, Carabobo y Cojedes. 1926.

II. Falcón, Guárico, Lara, Mérida, Miranda y Monagas. 1926.

III. Nueva Esparta, Portuguesa, Sucre, Táchira, Trujillo, Yaracuy, Zamora, Zulia, Territorio Amazonas, Territorio Delta Amacuro y Distrito Federal. 1926.

IV. Resúmenes generales de la república, el de los venezolanos residenciados en el exterior y el comparativo entre los censos de 1920 y 1926. 1926.

CENSUS OF 1936

Dirección general de estadística.

Sexto censo nacional, 1936. Caracas, Tip. Americana, 1939–40. 3 vol. HA 1091.A57 1936

Primer volumen. Distrito Federal, 1939. 120 pp.

Segundo volumen. Estados Anzoátegui, Apure, Aragua, Bolívar, Carabobo, Cojedes, Falcón, Guárico, Lara y Mérida. 1939. 462 pp.

Tercer volumen. Estados Miranda, Monagas, Nueva Esparta, Portuguesa, Sucre, Táchira, Trujillo, Yaracuy, Zamora y Zulia; Territorio Federal Amazonas, Territorio Federal Delta Amacuro, Dependencias Federales y los Resúmenes generales de población de todas las entidades de la república. 1940. 572 pp.

There are two parts in the provincial tables. The first covers age, sex, marital status, education, fertility, native population, aliens, and rural-urban residence for the entire Province. The second gives age, sex, and marital status of the population of districts.

---

[4] There is no evidence that this census was ever taken. Population statistics in the *Anuario estadístico* for later years are estimates based on rates of natural increase and migration.

Resumen general del sexto censo de población, 26 de diciembre de 1936. Ley de censo nacional del 6 de julio de 1936 y decreto reglamentario del 11 de agosto del mismo año . . . Caracas, Tip. Garrido, 1938. 68 pp. HA 1091.A58 1936

CENSUS OF 1941

Boggio, Juan M.

Breves comentarios sobre los resultados preliminares del séptimo censo nacional de población. Venezuela, *Revista de fomento 4 (47)77–81.* April–June, 1942. Govt. Publ. R. R.

OTHER NATIONAL CENSUSES

CENSUS OF AGRICULTURE, 1937

Dirección general de estadística.

Censos agrícola y pecuario, 1937. Caracas, Lit. y tip. Casa de especialidades, 1939–41. 23 vol. HD 1911.A45 1937

Censo industrial, comercial y empresas que prestan servicios, 1936. Ed. oficial. Caracas, Tip. Carrido, 1938–1941. 21 vol. HA 1091.A55 1936

CURRENT NATIONAL VITAL STATISTICS

(Including Population Estimates)

Dirección general de estadística.

Anuario estadístico de Venezuela, 1940. Caracas, Tipografía Venezuela, 1941. vi, 710 pp. HA 1091.A42

This yearbook, the second since the reorganization of Venezuelan statistics, presents data for the years 1939 and 1940. The second part, "Estado y movimiento de la población," pp. 21–147, gives summary data from the Census of 1936. Vital statistics include marriages, births, deaths and stillbirths by states by months, and migration, 1940. Numbers of marriages, births and deaths are also given by sex for minor civil divisions, 1940 and 1939. International statistics are then presented, based on data from the *Statistical Yearbook of the League,* with modifications of the rates for Venezuela.

Other sections of this *Anuario* present summary data from the agricultural, industrial and commercial censuses.

Boletín mensual de estadística, junio, julio y agosto de 1941. Año I (6–8), June–Aug., 1941. Caracas, 1942. Govt. Publ. R. R.

Sección III, "Sección de estadísticas demográficas," presents vital statistics for the second quarter of 1941, with comparative summaries for 1936–1941. Migration statistics are also included for the second quarter of 1941.

Ministerio de sanidad y asistencia social. Dirección de salubridad pública. División de epidemiología y estadística vital.

Relación anual de la sección de estadística vital, año 1939. Caracas, Lit. y tip. del comercio, 1940. 322 pp. Pan Am. San. Bur.

The first section, "Población," gives some population estimates for selected areas. There are estimates for the city of Caracas for the years 1934–1939, and an analysis of the post-censal increase. The section, "Natalidad," presents detailed statistics for selected areas. There are estimates for the city of Caracas for the years 1934–1939, and number of births by sex and legitimacy status for the provinces. The section, "Mortinatalidad," presents statistics on stillbirths in Caracas, 1939. The section, "Mortalidad," includes statistics on general, maternal and infant mortality, and deaths from selected causes. An appendix describes the organization of the Vital Statistics Branch and reproduces the forms employed in births and death registration. This is the first issue of a contemplated annual publication.

Part II

AMERICAN SECTIONS OF THE BRITISH COMMONWEALTH OF NATIONS

CANADA

Historical

The earliest recorded population counts in Canada date back to the founding of Port Royal in 1605 and Quebec in 1608. Estimates of the population in New France from 1608 to 1631 were based on yearly arrivals and departures, recorded vital statistics, and the memoirs and works of Champlain, Leclerq, Sagard, and others.[1] The earliest census, taken in New France in 1666, secured information on age, sex, occupation, and conjugal and family conditions. Fifteen additional censuses were taken prior to the end of the French regime in 1763, as well as seven for Nova Scotia and one for Prince Edward Island. In addition, 30 estimates, based primarily on local authority, cover the years 1665–1763.

Censuses were taken less frequently during the British colonial period, information being secured primarily from the reports of the colonial governors. Censuses of Canada were taken in 1765, 1784, and 1790. Upper Canada had annual censuses from 1824 to 1842, while Lower Canada had censuses in 1825, 1831, and 1834. After 1831 periodic censuses were taken of the western settlements.[2]

After Upper and Lower Canada were united, an act of 1841 provided for a census of all the provinces in 1842 and every fifth year thereafter. However, the census of Upper Canada was taken in 1842, that of Lower Canada in 1844. The census of Upper Canada of 1848 was the result of a similar act in 1847. An act passed in August 1851 provided that a census be taken in January 1852, in 1861, and every tenth year thereafter. Nova Scotia, New Brunswick, and Prince Edward Island also had censuses in 1851, so that actual national censuses date from 1851, although the first census after confederation was that of 1871.[3]

A complete listing of all counts and censuses of Canada from the earliest period to the present is given in the first volume of the Census of 1931. The Dominion Bureau of Statistics plans to publish a historical volume which will include all censuses taken since the founding of Port Royal, with a critique of the accuracy of each.[4]

An outstanding aspect of Canadian census development is the census of the Prairie Provinces. This regional census originated in the censuses of Manitoba and the Northwest Territories in 1886, which were followed by another census of Manitoba in 1896. The Census and Statistics Act was amended in 1905 to provide for a census of Manitoba, Saskatchewan, and Alberta in 1906 and every tenth year thereafter.[5] "The primary purpose of the census is to fix a basis for

[1] Dominion Bureau of Statistics. *Seventh Census of Canada, 1931.* Vol. I. Ottawa, J. O. Patenaude, 1936. p. 133. See also: *A travers les registres* de l'Abbé Cyprien Tanguay. Librairie St. Joseph, 1886.

[2] Dominion Bureau of Statistics. *Census of Canada, 1921.* Vol. I. Ottawa, A. A. Acland, 1924. pp. x–xi.

[3] Dominion Bureau of Statistics. *The Dominion Bureau of Statistics, its origin, constitutions, and organizations.* Ottawa, J. O. Patenaude, 1935. pp. 12–13.

[4] Dominion Bureau of Statistics. *Seventh Census of Canada, 1931.* Vol. I, p. 133.

[5] Dominion Bureau of Statistics. *Census of the Prairie Provinces, 1936.* Vol. I. Ottawa, J. O. Patenaude, 1938. p.v.

the per capita allowance payable to the provinces by the Dominion Government." More frequent censuses in these western provinces are therefore considered essential, since conditions change more rapidly than in the rest of the Dominion.

EARLY ESTIMATES AND CENSUSES

Department of Agriculture.

Censuses of Canada, 1665 to 1871. *Census of Canada, 1871.* Vol. IV. Ottawa, I. B. Taylor, 1876. lxxxv, 422 pp. HA 741.1871

The introduction presents a chronological list of the censuses and statements of population from the founding of Port Royal in 1605 to 1871. The major portion of the volume consists of summaries of censuses taken at different periods in and for the various territories of British North America. Ninety-eight official documents are summarized: 25 for Quebec, 22 for Ontario, 16 for Nova Scotia, 10 for Manitoba, 8 for New Brunswick, 6 for Prince Edward Island, and 1 for British Columbia. "The tables of this volume contain the information afforded by official documents, manuscript or printed, preserved in libraries and amongst the Public Archives, but classified and arranged uniformly so as to be easily consulted."

Dominion Bureau of Statistics.

Seventh Census of Canada, 1931. I. Population. Summary. Ottawa, J. O. Patenaude, I. S. O., 1936. 1520 pp.

All counts and censuses from the earliest period to the present are listed, pp. 133–153.

NATIONAL CENSUSES

CENSUS OF 1851

Board of Registration and Statistics.

Census of the Canadas, 1851–2. Quebec, John Lovell, 1853–1855. HA 741.1851

Vol. I. Personal census. 1853. xliii, 586 pp.

Vol. II. Agricultural produce, mills, manufactories, houses, schools, public buildings, places of worship, etc. 1855. 474 pp.

CENSUS OF 1861

Board of Registration and Statistics.

Census of the Canadas, 1860–61. Quebec, S. B. Foote, 1863–1864. HA 741.1861

Vol. I. Personal census. 1863. 590 pp.

Vol. II. Agricultural produce, mills, manufactories, houses, schools, public buildings, place of worship, etc. 1864. 363 pp.

CENSUS OF 1871

Department of Agriculture.

Census of Canada, 1870–71. Ottawa, I. B. Taylor, 1873–1878. HA 741.1871.

Vol. I. 1873. xxvii, 455 pp.
Vol. II. 1873. lx, 463 pp.
Vol. III. 1875. xii, 479 pp.
Vol. IV. Statistics of Canada. 1876. lxxxv, 422 pp.
Vol. V. Statistics of Canada. 1878. xlvii, 479 pp.

CENSUS OF 1881

Department of Agriculture.

Census of Canada, 1880–81. Ottawa, MacLean, Roger and Co., 1882–1885. HA 741.1881.

Vol. I. 1882. xv, 443 pp.
Vol. II. 1884. ix, 467 pp.
Vol. III. 1883. xi, 537 pp.
Vol. IV. Final report. 1885. xxxi, 181 pp.

CENSUS OF 1891

Department of Agriculture.
Census of Canada, 1890–91. Ottawa, S. E. Dawson, 1893–1897. HA 741. 1891
Vol. I. 1893. xxi, 403 pp.
Vol. II. 1893. xi, 361 pp.
Vol. III. 1894. vii, 401 pp.
Vol. IV. 1897. xi, 570 pp.

CENSUS OF 1901

Department of Agriculture.
Fourth Census of Canada, 1901. Ottawa, S. E. Dawson, 1902–1906.
HA 741.1901
Vol. I. Population. 1902. xxiii, xxv, 513 pp.
Vol. II. Natural products. 1904. xc, xcvii, 444 pp.
Vol. III. Manufactures. 1905. lxxvii, lxxxv, 384 pp.
Vol. IV. Vital statistics, school attendance, educational status, dwellings and families, institutions, churches and schools, electoral districts and representation. 1906. vii, vi, 467 pp.

CENSUS OF 1911

Census and Statistics Office.
Fifth Census of Canada, 1911. Ottawa, C. H. Parmalee, 1912–1915.
HA 741.1911b
Vol. I. Areas and population by provinces, districts and subdistricts. 1912. viii, xi, 523 pp.
Vol. II. Religions, origins, birthplace, citizenship, literacy and infirmities, by provinces, districts and subdistricts. 1913. xvii, xviii, 654 pp.
Vol. III. Manufactures for 1910 as enumerated in June, 1911. 1913. xv, xvii, 432 pp.
Vol. IV. Agriculture. 1914. xcv, cii, 428 pp.
Vol. V. Forest, fishery, fur and mineral production. 1915. l, lii, 171 pp.
Vol. VI. Occupations of the people. 1915. xxxi, xxxi, 469 pp.

CENSUS OF 1921

Dominion Bureau of Statistics.
Sixth Census of Canada, 1921. Ottawa, F. A. Acland, 1924–1929.
HA 741.1921.A3
Vol. I. Population. Number, sex and distribution—racial origins—religions. 1924. xcvii, 859 pp.
Vol. II. Population. Age, conjugal condition, birthplace, immigration, citizenship, language, educational status, school attendance, blindness and deaf mutism. 1925. xlviii, 776 pp.
Vol. III. Population. Dwellings, families, conjugal condition of family head, children, orphanhood, wage earners. 1927. 551 pp.
Vol. IV. Occupations. 1929. cxlvii, 837 pp.
Vol. V. Agriculture. 1925. cxviii, 787 pp.

CENSUS OF 1931

Dominion Bureau of Statistics.
Seventh Census of Canada, 1931. Ottawa, J. O. Patenaude, I. S. O., 1933—1936.
HA 741.1931.A43
Vol. I. Population. Summary. 1936. 1520 pp. [The growth of population in Canada, rural and urban distribution, age, conjugal condition, birthplace, year of immigration, nativity of parents, racial origin, religions, official language, mother tongue, nationality, illiteracy, school attendance, the Canadian born, the immigrant population, the gainfully occupied, unemployment, families and earnings, housing and rentals, farm population and farm workers, institutions, the blind and deaf, census of merchandising and service establishments.]
Vol. II. Population by areas. 1933. xv, 939 pp. [Population by counties or census divisions, rural and urban, sex, ages, conjugal condition, racial origin, religion, birthplace, immigration, citizenship, language spoken, mother tongue, literacy, school attendance.]
Vol. III. Ages of the people classified by sex, conjugal condition, racial origin, religion, birthplace, language, literacy, school attendance, year of immigration, naturalization, etc. 1935. xvi, 1009 pp

Vol. IV. Cross-classification. 1934. xx, 1413 pp. [Birthplace, Canadian born, immigration, citizenship, racial origin, language spoken and mother tongue, literacy, illiteracy, school attendance.]

Vol. V. Earnings of wage-earners, dwellings, households, families, blind and deaf-mutes. 1935. xivii, 1730 pp.

Vol. VI. Unemployment. 1934. xxvii, 1319 pp. [Unemployment on June 1, 1931, unemployment during the period June 1, 1930 to June 1, 1931, Appendix I, Unemployment in urban centers of 1,000 population and over. Appendix II, Unemployment in Canada, Census 1921.]

Vol. VII. Occupations and industries. 1936. xxvii, 1007 pp. [The appendix gives figures for 1891, 1901 and 1921.]

Vol. VIII. Agriculture, 1936. ccxxii, 838 pp.

Vol. IX. Institutions. 1935. iii, 495 pp. [Hospitals for the sick; mental and neurological institutions; charitable and benevolent institutions; penitentiaries, corrective and reformative institutions.]

Vol. X. Merchandising and service establishments—Part I. Retail merchandise trade. Summary for Canada and statistics for provinces, cities, towns, and counties or census divisions. 1934. lxxxiii, 1077 pp.

Vol. XI. Merchandising and service establishments—Part II. Retail services—wholesale trade—retail chains—hotels. Distribution of sales of manufacturers. 1934. xii, 1298 pp.

Vol. XII–XIII. Census monographs. [Nos. 2–9, 11, and 13 were published as separates by the latter part of 1942, and the remainder were in course of preparation.]

1. Population growth.
2. Age distribution of the Canadian people.
3. Fertility of the population of Canada.
4. Racial origins and nativity of the Canadian people.
5. Illiteracy and school attendance in Canada.
6. Rural and urban composition of the Canadian people.
7. The Canadian family.
8. Housing and rentals in Canada.
9. Dependency of youth.
10. Occupational structure of the Canadian people.
11. Unemployment.
12. Population basis of agriculture.
13. Canadian life tables, 1931.

Instructions to commissioners and enumerators. Ottawa, 1931. 205 pp.

Canadian abridged life tables, 1871, 1881, 1921, 1931. Ottawa, 1939.

Eight abridged life tables are presented for males and females covering the years 1871, 1881, 1921 (Registration area), and 1931 (Canada, excluding Yukon and Northwest Territories). For 1871 and 1881 the data cover only the provinces of Nova Scotia, New Brunswick, Quebec, and Ontario.

CENSUS OF 1941

Dominion Bureau of Statistics.

Eighth Census of Canada, 1941. Population. Ottawa, 1942–1943.

The following series of bulletins for Canada and each province present final population figures of the 1941 census:

A–1. Counties and census divisions. [By sex, for rural and urban subdivisions.]

A–2. Conjugal condition. [Population by conjugal condition and sex for rural and urban subdivisions.]

A–3. Age. [Population by quinquennial age groups for each sex and the total for rural and urban subdivisions.]

A–4. Racial origin. [Population by racial origin by sex for rural-urban subdivisions.]

A–5. Religion. [Population by 23 religious denominations by sex for rural and urban subdivisions.]

A–6. Birthplace. [Place of birth by province within Canada and by country abroad by sex for rural and urban subdivisions.]

A–7. Immigration and citizenship. [Canadian born, immigrant population by period of immigration, naturalized by period of naturalization, and aliens by country of allegiance, by sex, for rural and urban subdivisions.]

A–8. School attendance and years of schooling. [Population of school age, school attendance, and years of schooling by age, by sex, for rural and urban subdivisions.]
A–9. Language and mother tongue. [Population by official language and mother tongue by sex for rural and urban subdivisions.]

Occupations and earnings. Bulletin No. 1. Ottawa, April 28, 1942 to . . . .
Wage earners, 14 years and over, are classified by sex, showing the number and percent distribution of earning groups, for Canada and economic regions, and also for rural and urban areas.

Housing. Ottawa, Dec. 8, 1941 to . . .
Nos. 1–27 give preliminary data on housing for cities of 30,000 population and over. No. 28 is a summary bulletin. Nos. 29–31 are as follows: Crowding in Canadian cities of 30,000 population and over; Average earnings per person and rooms per person among wage-earner private families; and Canadian farm homes and households.

Households, occupations and earnings.
Preliminary reports are being issued on the basis of a ten per cent sample.

Agriculture.
Preliminary reports are being issued on agriculture and the farm population.

REGIONAL CENSUSES

Censuses of the Prairie Provinces

CENSUSES OF 1884–1886

Department of Agriculture.

Census of the three provisional districts of the Northwest Territories: Assiniboia, Saskatchewan and Alberta, 1884–5. Ottawa, MacLean, Roger and Co., 1886. 97 pp. HA 747.A38.1886

Census of Manitoba, 1885–6. Ottawa, MacLean, Roger and Co., 1887 [not numbered]. HA 747. M5 1886a

CENSUS OF 1906

Department of Agriculture.

Census of population and agriculture of the Northwest Provinces: Manitoba, Saskatchewan, Alberta, 1906. Ottawa, S. E. Dawson, 1907. xxxii, 160, xxxiii, 4 pp. HA 747. A4. 1906

CENSUS OF 1916

Census and Statistics Office.

Census of the Prairie Provinces. Population and agriculture: Manitoba, Saskatchewan, Alberta, 1916. Ottawa, J. de Labroquerie Taché, 1918. lxiv, lxvi, 356 pp. HA 747. A4 1916

CENSUS OF 1926

Dominion Bureau of Statistics.

Census of the Prairie Provinces, 1926. Ottawa, F. A. Acland, 1927. HA 747. A4 1926

Census of Alberta, 1926. Population and agriculture, 269 pp.
Census of Manitoba, 1926. Population and agriculture, 205 pp.
Census of Saskatchewan, 1926. Population and agriculture, 299 pp.

CENSUS OF 1936

Dominion Bureau of Statistics.

Census of the Prairie Provinces, 1936. Ottawa, J. O. Patenaude, 1938. HA 747. A4 1936

Vol. I. Population and agriculture. cxi, 1276 pp. [Part I. Population. Age, conjugal condition, birthplace, racial origin, immigrant population, citizenship, naturalization, language and mother tongue, years at school, literacy, school attendance. Part II. Agriculture. Farm population, farm workers and weeks of hired labour, age of farm operator, years a farmer and years on present farm, birthplace of farm operator, racial origin of farm operator, immigrant farm operators and period of residence in Canada.]

Vol. II. Occupation, unemployment, earnings and employment, households and families. lxvii, 1357 pp. [Part I. Gainfully occupied. Occupation, age, conjugal condition, birthplace, period of arrival, racial origin, status, years at school, industry, retired. Part II. Wage-earners. Unemployment on June 1, 1936, earnings and employment during census year ended June 1, 1936. Part III. Buildings, dwellings, and households. Buildings, all households, normal households, wage-earner households, relief households, value of home, monthly rent, rooms occupied, kind of dwelling, size of household, families in household, lodgers, birthplace of head, racial origin of head, occupation of head, earnings of head. Part IV. Families. All families, normal families, wage-earner families, relief families, age of head, conjugal condition of head, female heads, birthplace of head, racial origin of head, years at school of head, occupation of head, earnings of head.]

CURRENT NATIONAL VITAL STATISTICS

Dominion Bureau of Statistics. Vital Statistics.

Vital Statistics, 1940. Twentieth annual report. Ottawa, Edmond Cloutier, 1942. 450 pp. Govt. Publ. R. R.

Comparative and analytical tables on population, births, infant and general mortality, causes of death, and marriages by provinces and minor civil divisions. Information is given for the years 1921–1940. There are detailed tables for 1940.

Preliminary annual report, vital statistics of Canada, exclusive of Yukon and the Northwest Territories, 1941. Final figures, with rates computed on final census population of 11,506,655. Ottawa, 1943. 28 pp.

Preliminary report on births, deaths, and marriages in the second quarter of 1942. Ottawa, 1943. 6 pp. Mimeo. Bilingual. Govt. Publ. R. R.

Births, deaths, and marriages, by province; causes of death for selected causes; maternal mortality from selected causes, compared with fourth quarter of 1940; deaths at all ages from specified causes by provinces; résumé of births, deaths, and marriages by provinces, 1941; deaths at all ages for specified causes, by provinces.

Registration of births, deaths, and marriages, March, 1943. Enregistrement des naissances, décès et mariages, Mars, 1943. Ottawa, 1943. 2 pp. Mimeo. Govt. Publ. R. R.

This monthly report gives numbers of births, deaths, and marriages in cities, towns, and villages with populations of 10,000 and over, based on a count of the registrations filed during the month.

OTHER CURRENT NATIONAL POPULATION STATISTICS

Dominion Bureau of Statistics.

The Canada Year Book, 1942. Ottawa, King's Printer, 1942. 1030 pp. HA 744. S81

Chapter IV. Population, pp. 82–99. Section 1 consists of census statistics of the general population since 1871. It includes total population, percentage distribution, and percentage change by provinces and territories; area and density of population by provinces, counties, and census divisions (1941), density in various countries of the world, and a summary of births, deaths, natural increase and immigration, with estimated populations as of June 1, 1921–40. Section 8 gives rural and urban distribution, 1871–1931. Section 10 gives the population of cities and towns of over 5,000 inhabitants, 1871–1941. Section 17 gives estimates of the population for intercensal years. Section 18 is a discussion of the National Registration of 1940. Some additional material on population in the 1941 census is presented in Appendix III.

Chapter V, pp. 100–150, consists of five sections devoted respectively to marriages, births, deaths, natural increase, and vital statistics of the Yukon and Northwest Territories.

Chapter VI. Immigration and colonization, pp. 151–170, presents statistics on the growth of immigration, sex and conjugal condition, languages and racial origin, countries of birth and nationalities, destinations and occupations, rejections; and juvenile, refugee and oriental immigration. There are sections on emigration and colonization activities.

Appendix III, pp. 1004–1006, contains a table on sex distribution of the population, by provinces, 1871–1941.

Census figures for 1941 are preliminary.

NEWFOUNDLAND AND LABRADOR

Historical

Population information for Newfoundland has been collected at fairly short intervals since the founding of St. John's in 1613. There are several estimates for the next 67 years, the last of these being for 1680. The first census, taken in 1687, covered only French population and agriculture. Figures for the British population from 1714 through 1792 are included in the Newfoundland Report 1793, Appendix 6. Since no information is given as to the method of taking any of these counts, it is impossible to estimate their accuracy.

The first complete census of Newfoundland was taken in 1845, and censuses of Newfoundland and Labrador have been taken regularly since 1857. A complete list of censuses and estimates was published in Volume 1 of the Tenth Census.[1] Since Newfoundland has no regular census or statistics office of its own, the bibliographical material was supplied by the Dominion Bureau of Statistics of Canada.

The Census of 1935 was taken by the Department of Public Health and Welfare, which does not appear to have published any subsequent material on demography or vital statistics. Vital statistics are published annually by the Registrar General's Office.

EARLY CENSUSES

Canada. Department of Agriculture.

Censuses of Canada, 1665 to 1871. Census of Canada, 1871, Vol. IV. Ottawa, I. B. Taylor, 1876. lxxxv, 422 pp. HA 741.1871

The major portion of this volume consists of summaries of censuses taken at different periods in and for the various territories of British North America. The censuses of Newfoundland for 1687, 1691, 1692, 1696, 1698, 1705, 1711, 1845, 1857, and 1769 are included.

Newfoundland. Department of Public Health and Welfare.

Tenth Census of Newfoundland and Labrador, 1935. Vol. 1, Appendix A. St. John's. Printed by the Evening Telegram, Ltd., 1937. HA 747.No.52 1935a

Appendix A contains a chronological list of the censuses of Newfoundland, compiled by the Census Branch of the Dominion Bureau of Statistics. The number of persons wintering in St. John's is reported for 1613 and 1622. Estimates are reported from secondary sources for 1654, 1671, 1673, and 1680. Censuses of the French population and agriculture are reported for 1687, 1693, and 1694; censuses of the French population only in 1704 and 1706. Various special censuses are reported for Plaisance. A census of English population and agriculture is also reported for 1696. Estimates from various sources are reported for many of the intervening years.

NATIONAL CENSUSES

CENSUS OF 1857

Newfoundland.

Abstract census and return of the population, etc., of Newfoundland. 1857. . . . E. D. Shea, 1857. 125 pp. HA 747.N53

---

[1] Department of Public Health and Welfare. *Tenth Census of Newfoundland and Labrador, 1935.* Vol. 1. Population by districts and settlements. St. John's, N. F., printed by the Evening Telegram, Ltd., 1937. Appendix, A, pp. III–VI.

CENSUSES, 1869, 1874, 1884

The reports of these censuses were not located.

CENSUS OF 1891

Colonial Secretary's Office.

Census of Newfoundland and Labrador, 1891. St. John's, N. F., J. W. Withers, 1893. HA 747.N52 1891

Table I. Population, sex, condition, denominations, professions, etc. 469 pp.

Table II. Fisheries, property, produce of land, livestock, etc. Table III. Mines, factories, etc. 445 pp.

CENSUS OF 1901

Colonial Secretary's Office.

Census of Newfoundland and Labrador, 1901. St. John's, N. F., J. W. Withers, 1903. HA 747.N52 1901

Table I. Population, sex, condition, denominations, professions, etc. xxx, 457 pp.

Table II. Fisheries, property, produce of land, livestock, etc. Table III. Mines, factories, etc. 501 pp.

CENSUS OF 1911

Colonial Secretary's Office.

Census of Newfoundland and Labrador, 1911. St. John's, N. F., J. W. Withers, 1914. HA 747.N52 1911

Table I. Population, sex, condition, denomination, profession, etc. xxxi, 505 pp.

Table II. Fisheries, ships and boats. 487 pp.

Table III. Church buildings, superior and board schools, charitable and other institutions, etc. 594 pp.

Table IV. Ice, dogs, animals and animal products, forest products. Table V. Mines and minerals, tanneries, breweries, foundries, bakeries, furniture, factories, and manufactures or industries not otherwise enumerated. 395 pp.

CENSUS OF 1921

Colonial Secretary's Office.

Census of Newfoundland and Labrador, 1921. St. John's, N. F., 1923. HA 742.N52 1921

Table I. Population, sex, condition, denomination, profession, etc. xxiii, 512 pp.

Table II. Fisheries, ships, and boats. xi, 493 pp.

Table III. Church buildings, superior and board schools, charitable and other institutions, etc. 589 pp.

Table IV. Ice, dogs, animals and animal products, forest products. Table V. Mines and minerals, tanneries, breweries, bakeries, furniture factories, etc. 451 pp.

CENSUS OF 1935

Department of Public Health and Welfare.

Tenth Census of Newfoundland and Labrador, 1935. St. John's, N. F., Printed by the Evening Telegram, Ltd., 1937. HA 747.N52 1935

Vol. I. Population by districts and settlements. 961, xxix pp. [Historical information for different years is given for the following topics: Population distribution; religion; sex and age distribution; conjugal condition; birthplace; nationality; literacy; school attendance; defectives; orphans under age of 15 years. Appendix A contains a chronological list of Newfoundland censuses.]

Vol. II. Part I. Families and dwellings, occupations, and earnings, buildings. Part II. Vessels, boats and gear. Occupied land and livestock. 675 pp. [Part I, Section I, Families and dwellings, gives the number of persons per house and per family; number of families and dwellings; monthly rent; and the value of owned dwellings, by districts. There are separate tables for St. John's on families and dwellings, and monthly rent and value of owned dwellings. Section II, Occupations, earnings and buildings, gives, for occupations and earnings, the number in various occupations, 1857–1921; number occupied and per cent of total population, 1857–1935; males and females in selected occupations and earnings reported; classification and class of worker; class of worker and number living on income;

and number reporting earnings by sex and industry groups. The section on buildings gives historical data from 1891 to 1935. Appendix A is a classification of industries.]

Part II, Section I, Fisheries, vessels, boats and gear, contains information on the number of people engaged in various parts of the fishing industry. Section II covers occupied land, produce, livestock, and products.

CURRENT NATIONAL VITAL STATISTICS

Registrar General's Office.

Annual report of the Registrar General of births, marriages, and deaths, for the year ended December 31, 1938. St. John's, N. F., Robinson and Co., Ltd., printers, 1940. HA 747.N5

This annual publication contains general summary tables on estimated population, natural increase, emigration and immigration, from 1907 to 1938; population and deaths at each age group; denominations; births, marriages and deaths by denominations and by districts. Births are given separately by sex and by district. Stillbirths and illegitimate births are given from 1929 to 1938. Deaths are given by age groups, districts and cause. There is a special section on deaths from tuberculosis, and a section on infant mortality.

COLONIES IN THE CARIBBEAN

Bahamas, Barbados, Bermuda, British Guiana, British Honduras, Falkland Islands, Jamaica and Dependencies, Leeward Islands, Trinidad and Tobago, Windward Islands.

Historical

Complete censuses were taken in few of the British colonies and possessions prior to the middle of the nineteenth century, although estimates and local enumerations of various types were made throughout the early period of European occupancy.[1] The decennial censuses which were taken in most of the colonies between 1861 and 1921 followed the general plan of the English census, with supplementary questions added according to local problems and interests. Financial stringency was responsible for the failure to take the 1931 census in many areas, while the difficulties of the war led to the cancelation of the 1941 censuses. There are only two recent censuses, those of Bermuda in 1939, and Jamaica in 1943.

Sparse populations, inadequate transportation and communication systems, ethnic and linguistic diversity, and other factors making the accurate enumeration of the population difficult have also hindered the development of adequate systems for the collection of vital statistics. Birth and death rates for British Guiana refer only to the coastal population, omitting the Indian and Negro inhabitants of the interior. The development of accurate vital statistics in British Honduras has been retarded by the mixed character of the population and the limited resources of the colonial government. The difficulties of inter-island communications have been one of the fundamental problems facing the registration officials of Bermuda. The factors facilitating or retarding the collection of complete and adequate statistics have varied from colony to colony, and from period to period. Hence the evaluation of these British colonial census and vital statistics cannot be made *a priori*, but only on the basis of an appraisal of specific data for the various colonies.[2]

Formal census publications constitute only a fraction of the materials available for the analysis of population trends and population problems in the British colonies. Publications of the Crown Agents for the Colonies, usually printed in

---

[1] For references to the literature see: Ragatz, Lowell J., Compiler. *A guide for the study of British Caribbean history, 1763-1834, including the abolition and emancipation movements.* Annual report of the American Historical Association, 1930. Vol. III. Washington, Govt. Printing Office, 1932. viii, 725 pp. Z 1502.B5.R221. Also: Pitman, Frank W. *The development of the British West Indies, 1700-1763.* New Haven, Yale University Press, 1937. xiv, 495 pp. See especially Appendix I, The population of the British West Indies, and Appendix II, An account of the number of Negroes imported and exported at Jamaica each year, 1702-1775. pp. 391-392.

The early estimates and counts for individual colonies are often summarized in later census reports. For instance, the results of the censuses of 1816 and 1820 and the triennial censuses to 1835 for British Honduras are reproduced as an appendix to the 1921 census: British Honduras. General Registry. Report on the census of 1921. Part I. Report. Part II. Tables . . . Belize, Govt. Printing Office, 1922. 141 pp.

[2] Kuczynski's study of colonial statistics led him to the conclusion that knowledge of colonial populations was utterly inadequate. Permanent census staffs exist in hardly any colony, the quality of a census depending primarily on the skill of the official directing the census and the funds put at his disposal. His conclusion is that the problem is not how to improve the existing colonial statistics, but rather, an absolutely new departure based on the actual problems and difficulties of the colonial areas, not on imitation of the census techniques devised for other countries. Kuczynski, Robert R. *Colonial population.* London, Oxford University Press, 1937. 101 pp. Introduction, pp. ix-xiv. HB 885.K8

the chief city of the colony, include not only the census reports, but also the reports of the Registrar-General, and the annual Blue Books. The publications of the Colonial Office of Great Britain include both annual reports on the social and economic progress of the various colonies, and such special studies as those of the Comptroller for Development and Welfare in the West Indies.[3] In addition, the reports of recent Royal Commissions have been concerned to greater or less degree with the social and economic aspects of population problems.[4]

Population statistics and studies are included at irregular intervals in other colonial publications. Census reports are sometimes published in the Official Gazettes of the individual colonies, or as supplements to them. Reports on population problems are occasionally included in the Gazettes.[5] Many reports of legislative councils or investigating committees also include discussions of the acute nature of the problems of overpopulation and possible remedial measures.[6]

The bibliography which follows lists first the general types of statistical sources available for all colonies. Censuses and current vital statistics are then listed separately for each colony. Census reports to which no specific citation is made were not located.

GENERAL SOURCES FOR POPULATION STATISTICS OF BRITISH COLONIES

Colony.

Blue Book. Colony of __________, 19__. Printed in colony. Separately catalogued.

Section 15 of each Blue Book is devoted to population and vital statistics. The form of the tables is standardized, although the amount of information actually presented varies from colony to colony. The tables call for data on the populations of divisions by sex and color, number of persons employed in agriculture, manufacturing and commerce, and the annual numbers of births,

---

[3] Great Britain. Colonial Office. *Development and welfare in the West Indies, 1940–1942.* Report by Sir Frank Stockdale, Comptroller for Development and Welfare in the West Indies. London, H. M. Stationery Office, 1943. 93 pp. Specific recommendations for development and welfare programs in the fields of public health, agriculture, labor, social welfare, and education are made in accordance with the provisions of the Colonial Development and Welfare Act of 1940. The report includes a summary of the background of the Act, as well as a general résumé for each of the individual colonies. Govt. Publ. R. R.

[4] See, for instance: Great Britain. West India Royal Commission. *West India Royal Commission, 1938–39. Recommendations.* Command 6174. London, H. M. Stationery Office, 1940. 30 pp. F 2131.G83. This report is a summary of the recommendations, to serve as a guide to action programs until such time as the report can be published in full. See also: International Labour Office. "A social programme for the British West Indies." *International Labour Review 41 (5): 517–528.* May, 1940. The program recommended here is based on the recommendations of the Royal Commission and the Labour Advisor to the Secretary of State for the Colonies. Extensive land settlement schemes with subsistence agriculture are recommended as the first approach to the solution of the problems of surplus population on estates. *See also:* Great Britain. Colonial Office. *Agriculture in the West Indies.* Colonial No. 182. London, H. M. Stationery Office, 1942.

Another recent outstanding report on the problems of the British West Indies is the following: Orde Brown, G. St. J. *Labour conditions in the West Indies.* Command 6070. London, H. M. Stationery Office, 1939. 216 pp. This report concerns labor conditions in the West Indies, British Guiana, British Honduras, the Bahamas, and Bermuda. The first section consists of a general report dealing with problems common to the majority of the colonies. The second consists of detailed special reports on each colony, including a discussion of the special circumstances of the colony and the possible solutions. Local potentialities for the employment of surplus labor were investigated in all areas.

[5] Supplement to the Leeward Islands Gazette of Thursday, the 22nd of February 1940, pp. 3–12, is devoted to "Statement by His Majesty's Government on colonial development policy, and including a statement on the recommendations of the West India Royal Commission." *Leeward Islands Gazette for the Year 1940,* Vol. lxviii. J 3.B7

[6] Barbados. Legislature. *Minutes of Council and Assembly, 1937–1938.* Report of the Commission appointed to enquire into the disturbances which took place in Barbados on the 27th of July and subsequent days. JI 37.H3

deaths, and marriages, together with data on infant mortality. The most recent issues available for the individual colonies are as follows:

Bahamas, 1939 — J 136.R2
Barbados, 1942 — J 137.R2
Bermuda, 1941 — J 131.R2
British Guiana, 1939 — J 146.R2
British Honduras, 1939 — J 144.R2
Falkland Islands, (1939?) — J 148.R2
Jamaica, 1938 — J 138.R2
Leeward Islands, including Dominica, 1939 — J 139.R2
Trinidad and Tobago, 1938 — F 2121.T7
Windward Islands:
  Grenada, 1938 — J 141.G7R2
  St. Lucia, 1938 — J 141.S4R2
  St. Vincent, 1939 — J 141.S8R2

Colony.

Official Gazette. Colony of ———, 19—. Printed in Colony. Separately catalogued.

Bahamas. Royal Gazette and Bahama Advertizer. Nassau, 1813–Oct., 1942. J 3.B2

Barbados. Official Gazette. Bridgetown, 1867–Nov., 1942. J 3.B3

Bermuda. Royal Gazette. Hamilton, 1907–Dec., 1942. J 3.B4

British Guiana. Official Gazette. Georgetown, 1841–Sept., 1942. J 3.B42

British Honduras. Government Gazette. Belize, 1898–Dec., 1942. J 3.B55

Jamaica, Jamaica Gazette. Kingston, 1781–Oct., 1943. J 3.B6

Also: Turks and Caicos Islands. Gazette. Grand Turk, 1914–1939. J 3.B65

Leeward Islands. Leeward Islands Gazette. 1904–Nov., 1942. J 3.B7

Trinidad. Trinidad Royal Gazette. Port-of-Spain, 1875–1940. J 3.B8

Windward Islands. Grenada. Government Gazette. St. George, 1900–Oct., 1942. J 3.B47

St. Lucia. St. Lucia Gazette. Castries, 1900–Nov., 1942. J 3.B9S3

St. Vincent. St. Vincent Government Gazette, 1879–1939. J 3.B9.S35

Great Britain. Colonial Office.

Colonial reports. Annual report on the social and economic progress of the people of ———, 19— Annual No. ——— London, H. M. Stationery Office, 19—. JV 33.G7.A4

Chapter III of each report usually includes an estimate of the total population and general vital statistics. The most recent report for each of the colonies is as follows:

Bahamas — Annual No. 1901, 1938.
Barbados — Annual No. 1898, 1938–39.
Bermuda — Annual No. 1899, 1938.
British Guiana — Annual No. 1926, 1938.
British Honduras — Annual No. 1894, 1938.
Falkland Islands — Annual No. 1888, 1938.
Jamaica — Annual No. 1896, 1938.
Cayman Islands — Annual No. 1872, 1937.
Turks and Caicos — Annual No. 1927, 1938.
Leeward Islands, including Dominica — Annual No. 1928, 1938.
Grenada — Annual No. 1923, 1938.
St. Lucia — Annual No. 1929, 1938.
St. Vincent — Annual No. 1933, 1938.
Trinidad and Tobago — Annual No. 1915, 1938.

BAHAMAS

CENSUSES, 1851–1931

Bahamas.

Report on the census of the Bahama Islands taken on the 5th April, 1891. Nassau, Nassau Guardian, 1891. 4 pp. DC

Report on the census of the Bahama Islands taken on the 14th April, 1901. Nassau, Nassau Guardian, 1901. 16 pp. HA 861.A5 1901

Report on the census of the Bahama Islands taken on the 2nd April, 1911. Nassau, Nassau Guardian, 1911. 5 pp. HA 861.A5 1911

Report on the census of the Bahama Islands taken on the 24th April, 1921. Nassau, Nassau Guardian, 1921. 4 pp. HA 861.A5 1921

Report on the census of the Bahama Islands taken on the 26th April, 1931. Nassau, Nassau Guardian, 1931. 4 pp. HA 861.A5 1931

Data for the individual districts include age, sex, education, birthplace, and occupational and marital status. Information on housing is included.

CURRENT VITAL STATISTICS

Bahamas. Medical Department.

Abridged medical and sanitary report for the year ending 31st December, 1941. Nassau, 1942. 8 pp. RA 194.B3B1

A summary of the vital statistics collected by the Registrar-General's Office is included.

Bahamas. Out Island Administration.

Reports for the year 1939. Nassau, 1940. 124 pp. J 136.T308

Population estimates, vital statistics, and internal migration are among the subjects covered in the individual reports for each of the inhabited islands.

BARBADOS

CENSUSES, 1851–1921

Barbados. Governor, 1866.

Report upon the population of Barbados, 1851–1871. Barbados, Barclay and Fraser, Printers to the Legislature, 1872. 25 pp. HA 865.A2 1871

A discussion of the variations in the distribution of the population during the period 1851–1871 is included, with data for 1871 in detail.

Census of Barbados, 1881–1891. Barbados, T. E. King and Co., Printers to the Legislature, 1891. 99 pp. DC

Comparative data for 1861 and 1871 are included with the detailed data for 1881–1891.

Census of Barbados, 1891–1911. Bridgetown, 1911. 44 pp. DC

Published as a supplement to the Official Gazette of Dec. 7, 1911.

Report on the census of Barbados, 1911–1921. Bridgetown, Advocate Co., Ltd., Printers to the Government, 1921. 115 pp. HA 865.A2 1911–21

The age, sex, marital status, occupation, birthplace, religion, and education of the population are covered, with additional information on housing and the employment of children. Comparative data for earlier censuses are included.

CURRENT VITAL STATISTICS

Barbados. Registration Office.

Report on the marriages, births and deaths . . . for the year ended 31st December, 1938. Supplement to Official Gazette, Aug. 31, 1939. Bridgetown, 1939. 12 pp.

The population is estimated by sex as of the end of 1938. Vital statistics are presented in detail for 1938, with general series for the earlier periods.

BERMUDA

CENSUSES, 1841–1939

Bermuda.

Census of the Bermudas or Somers Islands. A comparative statement for the years 1861, 1871, 1881 and 1891. Bermuda, 1891. 1 table. Army Medical Library

Census of the Bermudas or Somers Islands, 1891. Abstract of summaries of the several districts. Bermuda, 1892. 1 table. Army Medical Library

Census of the Bermudas or Somers Islands, 1901. Bermuda, 1902. 7 tables. Army Medical Library

Census of the Bermudas or Somers Islands, 1911. Bermuda, 1912. 8 tables. Army Medical Library

Bermuda Blue Book, 1924. London, Waterlow and Sons Ltd., 1926. 113 pp. J 131.R2

The Census of 1921 is summarized briefly, p. 57.

Bermuda. Census Commissioners.

Census of the Bermudas or Somers Islands, 1931. Separate sheets. No place or date of publication. DC

The following tables are included: 1. Abstract of summaries of the several districts, giving population by sex, race and age. 2. Marital condition by race, literacy, and sex; education and school houses; licensed houses, such as clubs, hotels, etc. 3. Religious profession and place of worship, by race. 4. Occupation. 5. Place of birth, the handicapped, etc. 6. Land cultivated and uncultivated, livestock, agricultural and horticultural produce, houses inhabited, etc. 7. Districts. Population, number employed in agriculture and commerce, place of birth, education, religion, land under tillage, agricultural products, horticultural products, houses. 8. Abstract of summaries of naval and military districts. Population by age, sex, and race, marital status, education, religion, place of birth, and acreage.

Bermuda Blue Book, 1939. Bermuda, Hamilton Press Co., 1939. 218 pp. J 131.R2

A summary of the data of the census of March 26, 1939, is included, p. 125.

CURRENT VITAL STATISTICS

Bermuda Islands. Medical and Health Department.

Report for the year 1941. Hamilton, Government Printer, 1942. 22 pp. RA 194.B4B12

Bermuda. Registrar General.

Report of the Registrar General for the year 1940. Bermuda Government Printer, 1941. 28 pp. Army Medical Library

Annual vital statistics are given for the period 1931–1940. The 1939 census distributions by sex and color are reproduced and estimates made as of the end of 1940.

BRITISH GUIANA

CENSUSES, 1861–1931

British Guiana. Census Board.

Results of the decennial census of the population of British Guiana taken on the 7th April, 1861. Demerara, Printed at the "Royal Gazette" Office, 1862. 26 pp. HA 1037.B7A5.1861

Results of the decennial census of the population of British Guiana, taken on the 3rd April, 1871. Demerara, Printed at the "Creole Office", 1872. iii, 55 pp. HA 1037.B7A5.1871

Results of the decennial census of the population of British Guiana taken on the 3rd April, 1881. Demerara, Printed at the "Argosy" Office, 1882. v, 55 pp. HA 1037.B7A5.1881

Preliminary report on the census, 1891. Georgetown, Demerara, C. K. Jardiné, 1891. 7 pp. HA 1037.B7A5.1891

The final report of this census was not located.

British Guiana. Census Commissioner's Office.

Report on the results of the census of the population, 1911. Georgetown, Demerara, Argosy Co., 1912. xxxv, 71 pp. HA 1037.B7A5.1911

Report on the results of the census of the population, 1921. Georgetown, Demerara, Argosy Co., 1922. xxxviii, 75 pp. HA 1037.B7A5.1921

Preliminary report on the census of the colony of British Guiana, 1931. Georgetown, Demerara, Argosy Co., 1931. 8 pp. HA 1037.B7A5.1931

Report on the results of the census of population, 1931. Georgetown, Demerara, Argosy Co., 1932. liii, 153 pp. DC

Subjects covered include the following: (1) Population—number, increase, and distribution; (2) Density of population; (3) Houses; (4) Sexes; (5) Ages; (6) Conditions as to marriage; (7) Birth places and races; (8) Occupations of the people; (9) Religious persuasion; (10) Literacy of the people in races; (11) Infirmities; (12) East Indian population; (13) Cost of Census; (14) Conclusion.

CURRENT VITAL STATISTICS

British Guiana. General Register Office.

Report of the Registrar General for the year 1940. Georgetown, Demerara, Argosy Co., 1941. 8, xxvii pp. HA 1037.B7A3

Estimated population by race; births and deaths by age; stillbirths; causes of death and infant mortality, by registration districts; and causes of death by age for the various races.

British Guiana. Medical Department.

Report of the Director of Medical Services for the year 1941. Georgetown, Argosy Co., 1942. 14 pp. RA 222.B7A3

Vital statistics are summarized in this report on public health.

BRITISH HONDURAS

CENSUSES, 1816–1931

[Reports of the censuses of 1816, 1820, 1823, 1826, 1829, 1832, 1835, 1861, 1871 and 1881 were not located. See Census of 1921 for reference to summary tables.]

British Honduras.

Results of census, April 6, 1891. Belize, 1891. DC

Report on the results of the census of the colony of British Honduras, taken on the 31st March 1901. Belize, Printed at the "Angelus" Office, 1901. 31 pp. Army Medical Library

Report on the result of the census of the colony of British Honduras, taken on the 2nd April 1911. Belize, Printed at the "Angelus" Office, 1912. 46 pp. Army Medical Library

British Honduras. General Registry.

Report on the census of 1921 . . . taken on the 24th April, 1921. . . . Belize, Govt. Print. Office. 1922. 2 vol. Part I, Report; Part II. Tables. 141 pp. HA 791.A5.1921

The census history of British Honduras is sketched briefly and summary data from the censuses of 1816, 1820, 1823, 1826, 1829, 1832, 1835, 1861, 1871, and 1881 are included.

Census of British Honduras, 1931. Belize, Printed by the Government Printer, 1933. 72 pp. HA 791.A5.1931

Age, sex, religion, birthplace and nationality, literacy, occupation, infirmities, and housing of the population are covered, with age and sex cross-classifications for many of the characteristics. A résumé of population trends from 1861 through 1931 is included.

CURRENT VITAL STATISTICS

British Honduras. General Registry.

Report on the vital statistics of British Honduras for the year 1939. Belize, Government Printer, 1940. 27 pp. Army Medical Library

Vital statistics are included for 1935–1939, with classifications by sex and race.

British Honduras. Medical Department.

Annual medical and sanitary report for the year ending 31st December, 1941. Belize, Government Printer, 1942. 11 pp. RA 191.B7A3

Summary vital statistics are included.

FALKLAND ISLANDS AND DEPENDENCIES

CENSUSES, 1881–1921

Falkland Islands.

Census of the Falkland Islands, 3rd April, 1881. Stanley, 1881 (?) 5 pp. HA 1105.F3A5 1881

Falkland Islands. Colonial Secretary, Supervisor of Census.

Report on census, 1901. Stanley, 26th September, 1901. 10 pp. HA 1105.F3A5 1901

Comparative data for 1881 and 1891 are included.

Falkland Islands. Registrar General and Supervisor of Census.

Report on census, 1911. Stanley, 1911. 12 pp. HA 1105.F3A5 1911

Report on census, 1921. London, Waterlow and Sons, Ltd., 1922. 12 pp. HA 869.F3A3 1921

Data are presented on distribution, and composition by sex, age, marital status, occupation and birthplace.

CURRENT VITAL STATISTICS

Falkland Islands.
Vital statistics for the Falkland Islands for the year ended 31st December, 1937. pp. 34–35 in: Falkland Islands Gazette, March 1, 1938. J 3.B45
The estimated population of the islands as of the end of 1937 is given by sex, with a separate estimate, by total only, for South Georgia. Marriages, births and deaths are given for three areas. Migration statistics are included.

Falkland Islands. Medical Department.
Annual medical and sanitary report for the year ended 31st December, 1939. Port Stanley, 1940. RA 237.F3A3
A summary report of vital statistics for 1939 is included.

JAMAICA AND DEPENDENCIES

Cayman Islands, Turks and Caicos Islands, Marant and Pedro Cays

CENSUSES, 1844–1921

Jamaica. Registrar General's Department.
Census of Jamaica and its dependencies taken on the 4th April, 1881. Kingston, Govt. Print. Establishment, 1882. 32 pp. HA 891.A4 1881
Comparative data for 1871 are included.
Census of Jamaica and its dependencies taken on the 6th April, 1891. Kingston, Govt. Printing Office, 1892. 80 pp. HA 891.A4 1891
Census of Jamaica and its dependencies taken on the 3d April, 1911. Kingston, Govt. Print. Office, 1912. 93 pp. HA 891.A4 1911
Census of Jamaica and its dependencies, taken on the 25th April, 1921 . . . Kingston, Govt. Print. Office, 1922. 76 pp. HA 891.A4 1921
Population data include color, age, sex, marital status, literacy, occupation, religion, and nationality. Special sub-classifications are given for the Indian population.

CURRENT VITAL STATISTICS

Jamaica. Medical Office.
Report for the year ended 31st December, 1941. Kingston, Government Printer, 1942. 4 pp. RA 194.J3A4

Jamaica. Registrar General's Department.
Annual report for the year ended 31st December, 1938. Kingston, Government Printer, 1940. 41 pp. HA 894.A3
The Annual General Report for Jamaica includes departmental reports, including that of the Registrar General. J 138.N3

THE LEEWARD ISLANDS

Antigua and Dependencies, Barbuda and Redonda, Monserrat, St. Christopher–Nevis and Dependency, Anguilla; the Virgin Islands; and Dominica (until Jan. 1, 1940)

CENSUSES, 1851–1921

[Few reports on the Censuses of 1851, 1861 and 1871 were located.]

Leeward Islands. Dominica. Registrar General's Office.
Census taken November 1871. Tables. Roseau, Contractor for the Public Printing, 1872. Unnumbered. HA 866.D6A5 1871
Census of Dominica, 1881. Roseau, A. T. Righton, 1881. Unnumbered. HA 866.D6A5 1881

Leeward Islands. Virgin Islands. Registrar General's Office.
Virgin Islands, census taken . . . on the 4th April, 1881. Tortola, 1881. 4 pp. HA 866.V5A5 1881

Leeward Islands. Colonial Secretary.
Leeward Islands. Census 1891, with tabular statements and report. Antigua, Printed by S. B. Laviscount, 1892? 4 pp. HA 866.A3 1891
Antigua, Barbuda, St. Kitts-Nevis, Anguilla, Dominica, Montserrat, Redonda and the Virgin Islands, with some materials from the Censuses of 1871, were included in the Census of 1891. Some data from the Censuses of 1871 are reproduced.

Leeward Islands.
Blue Book, 1901–2. Antigua, Government Contract Printer, 1902? Numbered in sections. J 139.R2
Statistics from the census of 1901 are reproduced. Areas covered are the same as those included in 1891.

Blue Book, 1911–12. No publisher's imprint. Numbered in sections. J 139.R2
Data from the Census of 1911 are included.

Leeward Islands. Antigua.
Report on the census of Antigua and its dependencies of Barbuda and Redonda, 1911. St. John, Govt. Print. Office, 1912. 23 pp. HA 866.A5A5 1911

Leeward Islands. Dominica. Registrar General's Office.
Census, 1911. Report dated 30th June, 1911. Roseau, Bulletin Office, 1911. 11 pp. HA 866.M6A5 1911

Leeward Islands. Montserrat. General Registry Office.
Montserrat census, 1911. Report. Bridgetown, Barbados, Advocate Co., Ltd., 1911. 14 pp. HA 866.M6A5 1911

Leeward Islands.
Blue Book, 1921. Antigua, Govt. Print. Office, 1921? Numbered in sections. J 139.R2

Leeward Islands. Antigua.
Report on the census of the island of Antigua and its dependencies, 1921. Antigua, Govt. Print. Office, 1922 (?) 6, 11 pp. Army Medical Library

Leeward Islands. Montserrat. General Registry Office.
Montserrat census report, 1921 . . . . London, Waterlow and Sons, 1921? 12 pp. HA 866.M6A5 1921

Leeward Islands. St. Kitts-Nevis. Registrar General.
St. Kitts-Nevis, census report, 1921. Roseau, Dominica, Bulletin Office, 1921? 5, 20 pp. Army Medical Library

CURRENT VITAL STATISTICS

Leeward Islands.
The Leeward Islands Gazette for the year 1940. Vol. LXVII. Antigua, Govt. Print. Office, 1941. J 3.B7
The supplements include the reports of the Registrar-Generals of Antigua and Montserrat, and of the Medical Officers of Antigua, Montserrat, and the Virgin Islands.

Leeward Islands. St. Christopher-Nevis.
Annual medical and sanitary report, 1939. St. Kitts, Printed at the St. Kitts Printery, 1940. Army Medical Library

TRINIDAD AND TOBAGO

POPULATION CENSUSES, 1851–1931

Trinidad. Government Statistician.
Census of the colony of Trinidad, 1891. Port-of-Spain, Govt. Print. Office, 1892. 32 pp. Army Medical Library
Summary data for 1851–1881 are included. Although Tobago was attached to Trinidad in 1888, it does not appear to have been included in this census.

Trinidad. Registrar-General's Department.
Census of the colony of Trinidad and Tobago, 1901. Port-of-Spain. Govt. Print. Office, 1903. 27 pp. Appendices. HA 867.A5 1901

Census of the colony of Trinidad and Tobago, 1911. Port-of-Spain. Govt. Print. Office, 1913. 82 pp. HA 867.A5 1911

Census of the colony of Trinidad and Tobago, 1921. Port-of-Spain, Govt. Print. Office, 1923. 189 pp. HA 867.A5 1921

Census of the colony of Trinidad and Tobago, 1931. Port-of-Spain, Govt. Print. Office, 1932. 46 pp. HA 867.A5 1931b
Sex, occupation, birthplace and nationality, religion, education, infirmities, and marital status of the population are given, frequently with cross-classification by age. There are separate tables for the East Indian population, floating population, and institutional inmates. Housing and livestock are also covered. Summary data are given for all census years since 1851.

CURRENT VITAL STATISTICS

Trinidad and Tobago. Registrar-General.

Vital statistics. Report of the Registrar General on the vital statistics of the colony for the year 1939. Trinidad and Tobago, Government Printer, 1940. 23 pp. HA 867.A3

Estimated populations are given as of the end of 1939 for principal towns and county divisions, with excess of births over deaths and immigration over emigration for the whole colony. Vital statistics are given in some detail, with separate data for the East Indian population. The report for the preceding year contained historical series.

Trinidad. Medical Department.

Health, medical and sanitary reports ... 1941. Port-of-Spain, Government Printer, 1943. 14 pp. RA 194.T7.A33

WINDWARD ISLANDS: DOMINICA

CENSUSES, 1851–1921

For Censuses of Dominica, see Leeward Islands

CURRENT VITAL STATISTICS

Dominica. General Register Office.

Annual report on the vital statistics of the colony of Dominica for the year 1941. Roseau, 1942. 6 pp. Govt. Publ. R. R.

Births are given by sex and legitimacy, deaths by age, sex, and cause, and marriages by religion, for fourteen districts.

WINDWARD ISLANDS: GRENADA

CENSUSES, 1851–1921

Grenada.

Report and general abstracts of the census of 1891, with graphic tables and notes thereon. St. George, Government Printer, 1891. 40 pp. HA 869.G7A3

Summary data are given from the censuses of 1851–1881. Carriacou was included.

Report and general abstracts of the census of 1901. St. George, Govt. Print. Office, 1902. 47 pp. HA 868.G7A5 1911

Report and general abstracts of the census of 1911. St. George, Govt. Print. Office, 1911. 32 pp. HA 868.G7A5 1911

Report on the 1921 census of Grenada. St. George, 1921 (?) 70 pp. HA 869.G7A3 1921

The population of districts are classified by sex, nationality, language, race, education, occupation, citizenship, and marital status. Housing data are also included.

CURRENT VITAL STATISTICS

Grenada. Registrar General.

Vital statistics. Registrar General's report for the year 1940. 12 pp. Army Medical Library

There are historical series for 1931–1940.

WINDWARD ISLANDS: SAINT LUCIA

CENSUSES, 1851–1921

Saint Lucia

Report on the census of the island of Saint Lucia, 1911. Castries, Govt. Print. Office, 1912. 68 pp. DC

Report on the census of the colony of Saint Lucia, 1921. Castries, Govt. Print. Office, 1921. iii, 61, 2 pp. HA 869.S4 1921

CURRENT VITAL STATISTICS

Saint Lucia. Registrar's Office.
Report of the Registrar of Civil Status for the year 1936. Castries, 1937. 8 pp.
HA 868.S3A3

St. Lucia. Medical and Sanitary Department.
Report for the year 1941. Castries, Govt. Print. Office, 1942. 10 pp.
RA 194.S3A3

WINDWARD ISLANDS: SAINT VINCENT

CENSUSES, 1851-1921

Saint Vincent.
Report and general abstracts of the census, 1911, with grafic tables . . . Kingstown, Govt. Print. Office, 1911. 26 pp. DC
That portion of the Grenadines dependent on Saint Vincent is covered.

Report and general abstracts of the census of 1921. Kingstown, Govt. Print. Office, 1922. 25 pp. HA 869.S4A5 1921

CURRENT VITAL STATISTICS

Saint Vincent.
Annual administration reports of the colony of Saint Vincent for the year 1940. Kingstown, Govt. Print. Office, 1941. 50 pp., plus agricultural report.
J 141 S8N15
Vital statistics are included, pp. 1–19.

Part III
AMERICAN COLONY OF DENMARK

GREENLAND

Historical

The historical censuses and vital statistics of Greenland have been developed as an integral part of the centralized statistical system of Denmark itself. A central statistical institution, the Dansk-Norsk-Tabelkontor, existed between 1797 and 1819. It was succeeded by a commission of members of various governmental departments, the Tabelkommission, which had charge of demographic statistics from 1834 to 1848. A statistical bureau was organized in 1850, although the financial resources were small. It achieved greater independence in 1896 and again in 1913, when it became known as the Statistical Department. [1]

In the early period both the taking of the census and the collection of vital statistics were the responsibilities of the clergy. In the census years of 1787 and 1801 the rural clergy enumerated the members of their congregations, while the magistrates had charge of the enumeration in the towns. The process of secularization was advanced in 1834, since, although the clergy supervised the enumeration, the actual collection of information was in charge of the school teachers. A house-to-house canvass replaced the earlier system of enumeration in the churches. Quinquennial censuses were taken between 1840 and 1860. In 1870 the local authorities, including the parish councils and the magistrates, executed the census, but the original name lists were sent to the central bureau for tabulation. Only decennial censuses were taken between 1860 and 1890, but beginning with 1901 and extending through 1940 there have been quinquennial enumerations.

The geographic, economic, and cultural differences between Greenland and Denmark are reflected in the relative completeness and the accuracy of the census statistics for the two areas.[2] Thirteen censuses have been taken in Greenland under Danish auspices. The first of these was taken during the early years of the nineteenth century, although it is variously cited as having been taken in 1801, 1802, and 1805.[3] The results of this census were presumably published with the returns of the census of 1834, in the Statistisk Tabelwaerk, aeldste Raekke, 6 Haefte. Later censuses were taken in 1840, 1845, 1855, 1860, and then decennially through 1930. A census of Denmark itself was taken in 1940, but it appears probable on the basis of available information that Greenland was not included.

---

[1] League of Nations. Health Organization. "The official vital statistics of the Scandinavian countries and the Baltic republics." *Statistical Handbooks Series*, No. 6. Geneva, 1926. 107 pp. (Denmark, pp. 56–65). HA 1466.L4 1926

[2] Greenland. Commission for the Direction of the Geological and Geographical Investigations in Greenland. Editors, Martin Vahl, George C. Andrup, Louis Bobe, and Adolf S. Jensen. Vol. I. The discovery of Greenland, exploration and nature of the country. Vol. II. The past and present population of Greenland. Vol. III. The colonization of Greenland and its history until 1929. Copenhagen, C. A. Reitzel. Vol. I., 1928, 575 pp.; Vol. II, 1928, 415 pp.; Vol. III, 1929, 468 pp. G 743.G73

See especially, in Vol. II, The Greenlanders of the present day, by Kai Birket-Smith; and On the Icelandic colonization of Greenland, by Finnur Jonsson; and in Vol. III, Sanitation and health conditions in Greenland, by Alfred Bertelsen.

[3] Danemark. Statens statistiske bureau. *Resume des principaux faits statistiques du Danemark, No. 1.* Copenhagen, Imprimerie de Bianco Luno, 1874. 81 pp. HA 1479

No. 1–2. Table I, Superficie, population d'après les recensements de 1801, 1840, 1860 et 1870, population calculée au 1er octobre, 1874. Included is a population for Greenland as of 1802. On the other hand, the report of the 1930 census states that this is the thirteenth census of Greenland, and gives data on the population in 1805, without reference to either 1801 or 1802. See: Denmark. Statistiske Departement. *Folketaellingen i Grønland den 1. oktober 1930.* Population de Gröenland au 1er octobre 1930. Statistiske Meddelelser, 4 Raekke, 87. Bind, 6 Haefte. København, Bianco Lunos Bogtrykkeri, 1932. 47 pp. HA 1473.B

The population statistics of Greenland have been published in the Danish statistical series. The earliest census located was that of 1880, but references to previous censuses have been taken from the census of 1930.[4] Summaries of census data were published in the *Sammendrag af Statistiske Oplysninger* from 1874 to 1893, and in the *Statistisk Aarbog* since 1896.[5] Current vital statistics also appear in abbreviated form in the Statistisk Aarbog.

CENSUSES

CENSUSES OF 1834–1860

Census of 1834. Statistisk Tabelvaerk, aeldste Raekke, 6. Haefte.
Census of 1840. Statistisk Tabelvaerk, aeldste Raekke, 10. Haefte.
Census of 1845. Statistisk Tabelvaerk, ny Raekke, 1. Bind.
Census of 1850. Meddelelser fra det statistiske Bureau, 4. Samling.
Census of 1855. Meddelelser fra det statistiske Bureau, 4. Samling.
Census of 1860. Statistiske Meddelelser, 2. Raekke, 4. Bind.

The original publications of these censuses were not located. Summary data and citations to sources were included in the report of the Census of 1930.

The returns of the Census of 1801 (or 1802 or 1805) presumably were included in the published report of the Census of 1834.

CENSUSES OF 1870–1880

Denmark. Statistiske Bureau.

Folketaellingen i Grønland den 1ste oktober 1880. pp. 83–128 in: Statistiske Meddelelser, 3 Raekke, 6. Bind. Kjøbenhavn, Bianco Lunos Kgl. Hof-Bogtrykkeri, 1883. 239 pp. HA 1473.B

Some comparative data are given for earlier censuses, particularly those of 1860 and 1870.

CENSUS OF 1890

Denmark. Statistiske Bureau.

Folketaellingen i Grønland den 1ste oktober 1890. pp. 273–321 in: Statistiske Meddelelser, 3 Raekke, 12. Bind. Kjøbenhavn, Bianco Lunos, Kgl.Hof-Bogtrykkeri (F. Breyer), 1892. 418 pp. HA 1473.B

CENSUS OF 1901

Denmark. Statistiske Bureau.

Folketaellingen i Grønland den 1ste oktober 1901. In: Statistiske Meddelelser, 4. Raekke, 14. Bind, Haefte v. Kjøbenhavn, Bianco Lunos Bogtrykkeri, 1904. 36 pp. DC

CENSUS OF 1911

Denmark. Statistiske Departement.

Folketaellingen i Grønland den 1. oktober 1911. In: Statistiske Meddelelser, 4. Raekke, en og fyrretyvende Bind, Haefte IV. Kjøbenhavn, Bianco Lunos Bogtrykkeri, 1913. 37 pp. HA 1473.B

CENSUS OF 1921

Denmark. Statistiske Departement.

Folketaellingen i Grønland den 1. oktober 1921. In: Statistiske Meddelelser, 4· Raekke, 66 Bind, Haefte 1–6. Kjøbenhavn, Bianco Lunos Bogtrykkeri, 1923. 40 pp. HA 1473.B

CENSUS OF 1930

Denmark. Statistiske Departement.

Folketaellingen i Grønland den 1. oktober 1930. Statistiske Meddelelser, 4. Raekke, 87. Bind, Haefte V. Kjøbenhavn, Bianco Lunos Bogtrykkeri, 1932. 47 pp. HA 1473.B

---

[4] Denmark. Statistiske Departement. *op. cit.*, p. 2.

[5] Denmark. Statistiske Departement. *Statistisk Aarbog.* Annuaire statistique. Kjøbenhavn, 1896–1941. HA 1477

Summary historical tables give the total and regional populations, 1834, 1860, 1880, 1890, 1901, 1911, 1921, and 1930; the sex ratio, and broad age distribution for 1860, 1890, 1911, 1921, and 1930. Data from the occupational census are summarized from the censuses of 1860, 1901, 1911, 1921, and 1930. Detailed tabulation for 1930 include the following:

Population in the various districts and communes, by sex, separately for natives and Europeans, together with number of places and number of households. *Ibid.* for health districts, electoral regions, etc. Native population of regions by quinquennial age groups by sex and marital status and by single years of age for all Greenland. European population by age, sex, and marital status, all Greenland. Population of districts by occupational composition. There is a special section for the district of Thuk, pp. 29–31.

CURRENT VITAL STATISTICS

Denmark. Statistiske Departement.

Folketaellingen i Grønland den 1. oktober 1930. Population de Gröenland au 1er octobre 1930. Statistiske Meddelelser, 4. Raekke, 87 Bind, 6. Haefte. Kjøbenhavn, Bianco Lunos Bogtrykkeri, 1932. 47 pp. HA 1473.B

The report of the 1930 census includes births and deaths for Greenland and each of the four regions, by single years, 1922–1930 inclusive. Distribution by months is given for the period as a whole, as are causes of death in the native population, and deaths by broad age groups. There is a brief summary of historical trends in mortality in Greenland, and comparisons with trends in Denmark itself, pp. 15–24.

Denmark. Statistiske Departement.

Statistisk Aarbog, 1941. Kjøbenhavn, Bianco Lunos Bogtrykkeri, 1941. xxxii, 307 pp. HA 1477.1941

Table 13, pts. 1–3, p. 230, includes total population figures from census data from 1805 to 1930; population by broad age groups, by sex and marital status in 1930; and marriages, births, and deaths for the years 1932–1938, by major regions.

Part IV

AMERICAN COLONIES OF FRANCE

FRENCH COLONIES

French Guiana, Guadeloupe, Martinique and Dependencies, St. Pierre and Miquelon.

Historical

The French colonies in the Americas consist of French Guiana, Guadeloupe, Martinique and Dependencies, and St. Pierre and Miquelon.[1] The literature on these colonies is profuse, but exact information concerning the extent and nature of their demographic statistics is quite elusive. The International Statistical Institute reports that there were censuses in Guadeloupe and Dependencies, French Guiana, Martinique, and St. Pierre and Miquelon in 1831, 1845 (except St. Pierre and Miquelon), 1852 (except French Guiana), 1861, 1876, 1906, 1911, 1921, 1926 (Martinique, 1927), 1931, and 1936. A census of French Guiana was presumably taken in 1829.[2] The difficulties of locating the original censuses is due in part to the fact that there has been no consistent plan of publication over any considerable period of time. In addition, French official publications tend to present tabular materials with no specific annotations as to source or method of compilation.

Early nineteenth century censuses, or estimates based on them, were published in detail by the *Ministère de la marine*, 1837, and the *Ministère des colonies*, 1842. Both census and vital statistics for the colonies were published in the *Statistiques coloniales*, issued by the *Ministère du commerce, de l'industrie et des colonies*, 1839–91, although issues for 1892–95 to 1898 included only commercial data. For the censuses of 1906 and 1911, special publications on population were issued by the *Ministère des colonies, Office colonial.*[3] Current population materials are carried occasionally in abbreviated form in the Bulletin, *Agence générale des colonies.*

The statistical publications of Metropolitan France have tended to contain increasing amount of material on the Colonial Empire. The *Annuaire statistique* carries summary data, and the first volume of each French census presents a résumé of census data for the individual colonies and possessions. Both of these sources present the colonial statistics without reference to the method of enumeration or the probable degree of completeness and accuracy of the returns.

[1] The Territory of Inini was created by a decree of June 6, 1930. For the governmental structure and internal organization of the various colonies, see: France. Ministère des colonies. *Annuaire . . .* 1934–1935. "Colonies françaises de l'Amerique," pp. 233–268. Paris, Charles-Lavauzelle et Cie., 1935. 79, 692, 142 pp. V33.F7.A2

[2] Institut international de statistique. *Aperçu de la démographie des divers pays du monde, 1929–1936.* La Haye, Office permanent de l'Institut . . . 1939. Table 3, pp. 44–50. HA 42.A65

[3] Ministère des colonies. Office colonial. *Statistiques de la population dans les colonies françaises pour l'année 1906, suivies du relevé de la superficie des colonies françaises.* Melun, Imprimerie administrative, 1909. 608 pp. HA 1228.A4 1904

Also, Ibid., 1911. Paris, Bureau de vente des publications coloniales officielles, 1914. 610 pp. A similar publication, issued for the 1921 census, was not located. HA 1228.A4 1911

There have been demographic publications by the individual colonies in the Americas, but these appear to have been issued only sporadically and in quite abbreviated form.[4]

The general paucity of published statistics reflects no lack of interest on the part of the French in the population problems and the vital trends in their Colonial Empire. French demographers have studied the available statistics with great care, attempting to infer from them the trends of growth of the native populations. They have also studied the problems involved in securing more adequate colonial vital statistics through the use of hospital records, experimental medical services, and other techniques. The general concensus of opinion among demographers who have made careful studies of the official population and vital statistics in the colonial possessions is that the censuses are so inaccurate and incomplete as to be of little analytical value for areas in most of Africa south of the Sahara, Oceania, and America.

Ulmer states that the censuses are often merely administrative estimates, made according to procedures that varied with the inclination of the administrator and local circumstances.[5] The dispersion and mobility of the population, the general lack of education, and the suspicion of the inhabitants, especially when a system of per capita taxes exists, makes any accurate count impossible. Ulmer cites the age distributions in the various French colonies as illustrative of the influence of fiscal considerations on demographic returns. In French West Africa the percentage of youth under 15 was returned as 21 percent in Senegal and the Ivory Coast, 23 percent in Guinea, and 37 percent in Dahomey. The explanation is that the age limit for the payment of the per capita tax is 16 in Dahomey, but 8 to 10 in the other colonies.[6]

Only Algeria and Indo-China among the French possessions have organizations whose major function is statistical work. In the majority of the colonies proper, census and other statistics are compiled on the basis of instructions sent out from the *Administration centrale des colonies*, and thus theoretically are uniform. Actually, the local administrators may or may not heed the instructions received. Even if they are convinced of the need of accurate and comprehensive statistics, they are faced by the almost insurmountable difficulties of attempting to secure replies to intricate census questionnaires from uneducated and dispersed natives, in a general atmosphere of suspicion and fear, and without the use of trained personnel.[7]

The difficulties involved in the use of French census statistics are nowhere better illustrated than in R. R. Kuczynski's note on the population of French Guiana and Inini Territory:

The published data are quite contradictory. *Résultats statistiques du recensement général*, Vol. I, Part I, p. 113, gives 28,310 Europeans and Assimilated (including 5,419 convicts), 1,000 Natives. *Bulletin de l'agence générale des colonies*, 1932, pp. 668–9, gives for French Guiana 22,169 (17,944 French; 2,934 Foreigners;

---

[4] cf.: France. Ministère des colonies. *Annuaire de la Guyane française, 1920–1922*. Cayenne, 1922. Commerce F2441.A3

Ibid. "La Guyane française." Melun, 1921 and 1923. Commerce HC 216.A4F8

Guadeloupe Island. *Annuaire de la Guadeloupe et dépendances, 1923*. Basse-Terre, 1923. Commerce F 2066.A4

Martinique. "Arreté portant publication des tableaux de la population de la colonie de la Martinique." *Journal officiel de la Martinique*, pp. 542–546. March 31, 1937. Commerce HA 920.M3A5 1937

Martinique. *Tableau statistique de la population*. Jan. 15, 1894. Commerce NW.MA2

[5] Ulmer, Henri. *Quelques données démographiques sur les colonies françaises*. Congrès international de la population (Paris, 1937) 6:110–127. Paris, Hermann et Cie., 1938. HB 849.I55 1938

[6] Ulmer, Henri. "La statistique dans les pays coloniaux." *Journal de la Société de statistique de Paris* *79:231–248*. 1937–1938.

[7] Ibid.

1,240 Liberated Convicts; 51 Relegated) and for Inini Territory 3,511 (983 French; 1,319 Foreigners; 441 Native Tribes; 767 Gold Seekers etc.). Colonies Autonomes, December, 1935, p. 173, gives the same totals, but adds: "These figures do not include the aboriginal Indians of the Interior who have fled from civilization." *Statesman's Year-Book*, 1936, p. 947, says that the figure of 22,169 (the author erroneously believes it to include Inini Territory) is "exclusive of the population of the penal settlement of Maroni, of the floating population of miners without any fixed abode, as also of officials, troops, and native tribes. . . . The military force consists of 310 officers and men of the Colonial Infantry. . . . In 1931 the penal population consisted of 5,954 men." *South American Handbook*, 1937, p. 367, states: "The population, inclusive of natives, is estimated at 25,679." Hubners *Geographisch-statistische Tabellen*, 1936, p. 228, gives as population 25,679 and "with wild natives," 45,679. *Almanach de Gotha*, 1937, p. 1052, gives as population 32,596. We have accepted the latter figure."[8]

Thus the French censuses seem to be primarily estimates of the local population made by the colonial administrators according to various methods in the different local areas. Vital statistics are equally inadequate for the majority of the colonies. The civil registration system is organized in only a small proportion of the territories, and even where it exists, it is often of such recent origin and so limited in scope that the numbers of births and deaths registered are totally inadequate as measures of the levels of fertility and mortality.[9]

CENSUSES AND ESTIMATES

CENSUSES OF FRANCE

Statistique générale de la France.

Résultats généraux du dénombrement de 1876. France, Algérie, Colonies. Paris, Imprimerie nationale, 1878. lxvii, 287 pp. HA 1219. 1876

Ch. III, Renseignements sur la population des colonies et autres possessions françaises, reports the number of persons in the French colonies and possessions on the basis of "the last statistics published by the Ministère de la marine et des colonies"; in another section, the data are referred to as "résultats généraux du dénombrement de la population, décembre, 1876." The detailed tables give, for Martinique and Dependencies, French Guiana, and St. Pierre and Miquelon the number of men and women, classified as infants, and single, married, or widowed. A total number is also given for the sedentary and the floating population.

Résultats statistiques du recensement général de la population effectué le 4 mars 1906. Tome I. Première partie. Introduction. Population légale ou de résidence habituelle pour la France entière. Paris, Imprimerie nationale, 1908. 126 pp. HA 1219. 1906

Appendice, Tableau II, gives the area, population, and density of population in the various colonies and protectorates as of the census of 1906. The data are credited to: Statistiques coloniales pour l'année 1906—population.

Succeeding censuses carry similar materials. For the last data available, those from the census of 1936, see: Résultats statistiques du recensement général de la population effectué le 8 mars 1936. Tome I. Première partie. Population légale ou de résidence habituelle. Appendice: Population des territoires françaises d'outremer et des pays étrangers. Paris, Imprimerie nationale, 1938. 114 pp.

ANNUAIRE STATISTIQUE, FRANCE

Statistique générale de la France.

Annuaire statistique, 1878–19—. HA1213.A4

The first issue, that for 1878, devoted a chapter to the colonies and possessions. Data entitled "Dénombrement et mouvement de la population, 1875" are given for the various colonies in the Americas, as well as those scattered throughout the remainder of the world. For a considerable period of time the even-numbered volumes contained a retrospective summary of international statistics, while the odd-numbered volumes contained statistics on the colonies. Since 1927 each

---

[8] Kuczynski, Robert R. *Colonial population*. Oxford University Press. 1937. pp 70–71. HB 885.K8

[9] Office international d'hygiène publique. *Essai de démographie des colonies françaises*. Supplément au Bulletin mensuel, xxx, 2. Feb., 1938. 154 pp. RA 421.03

volume has included both the annual tables for France and the colonies, and a retrospective résumé of the statistics of France and other countries.

COLONIAL REPORTS

Ministère du commerce, de l'industrie et des colonies.

Statistiques coloniales, 1839–1890. Paris, Imprimerie nationale. HA 1228.A2

This yearbook included both census and vital statistics from the date of its inception until 1891; issues between 1891 and 1898 included only commercial data. After 1898 separate volumes were issued covering the fields of commerce, finance, mining industry, navigation, population (occasionally on census years), and railroads. The contents of the early issues differed, both as to colonies included and the scope of the data. The last issue, that for 1890, gave for each of the American colonies the population in broad age groups by sex, vital statistics, and the number of immigrants by communes. Estimated year-to-year populations as presented in these yearbooks were secured by adding in the natural increase and the net migratory balance of the population of the previous year.

OTHER

Ministère de la marine.

Notices statistiques sur les colonies françaises. Première partie, Notice Préliminaire. Martinique, Guadeloupe et Dépendances. Seconde partie. Bourbon, Guyane Française. Quatrième et dernière partie. Possessions françaises à Madagascar. Iles Saint Pierre et Miquelon. Appendix. Paris, Imprimerie Royale, 1837–1840. JV 1825.A6. 1837

There is a preliminary notice concerning all the French colonies. The "Notice statistique sur la Martinique ' includes a brief history of the colony and chapters on topography, meteorology, population, legislation, finance, industry, commerce, etc. The chapter on population gives population as of Dec. 31, 1835, classified as free whites, free men of color, and slaves. The population is divided into broad age groups, separately by sex and for free and slave groups. There is also a distribution for the arrondissement of Fort Royal and St. Pierre, the cities, and the burgs of the rural areas. Vital statistics are given for 1835. Similar tabulations are presented for Guadeloupe and Dependencies.

The second part gives population and vital statistics similar to those for Guadeloupe for French Guiana for the year 1836 while the fourth part includes population and vital statistics for St. Pierre and Miquelon.

Ministère des colonies.

Tableaux et relevés de population, de cultures, de commerce, de navigation, etc., format pour l'année 1839, la suite des tableaux et relevés insérés dans les notices statistiques sur les colonies françaises. Paris, Imprimerie Royale, 1842. 141 pp. HA 1228.A2

Table 1 includes for each of the American colonies the total population in 1839, a classification into three broad age groups, rural-urban distribution, comparisons of the total with 1838, and the numbers of marriages, births, deaths, and natural increase separately for free and slave population. Similar volumes were issued in 1840 and 1841.

Ministère des colonies. Office colonial.

Statistiques de la population dans les colonies françaises pour l'année 1906 suivies du relevé de la superficie des colonies françaises. Melun, Imprimerie administrative, 1909. 608 pp. HA 1228.A4 1906

There is a summary table giving the 1906 census populations for each of the colonies. Sections on each colony, including the American ones, give more detailed data on sex, age groups, marital status, nationality, occupations, and migrants.

Statistiques de la population dans les colonies françaises pour l'année 1911 suivies du relevé de la superficie des colonies françaises. Paris, Bureau de vente des publications coloniales officielles. 1914. 610 pp. HA 1228.A4 1911

Similar data are presented for the 1911 census.

Recensement de la population des colonies françaises en 1921. Published in 1923. Not located.

CURRENT COLONIAL VITAL STATISTICS

Office international d'hygiène publique.

Essai de démographie des colonies françaises. Par Dr. Cazanove. Supplément au Bulletin mensuel 22 (8), août, 1930. Paris, 1930. 86 pp. RA 421.03

This compilation was based on documents supplied by the *Inspection générale du Service de santé des colonies*. Ch. VI summarizes the statistics for the Atlantic colonies. Dr. Cazanove stresses the fact that the civil authorities function primarily in cities, where the presence of hospitals and doctors artificially increases rates based on place of occurrence. The generally strict regulations requiring burial in official cemeteries, and the desire of the natives to obtain an exoneration of per capita taxes by a death certificate, also operate to increase the completeness of death registration. Births are much less likely to be declared, either because of negligence, or because of the rituals which impose a waiting period before the child can be named.

Essai de démographie des colonies françaises. Travail établi par les Drs. Martial et Beaudiment, d'après les documents de l'Inspection générale du Service de santé des colonies, et communiqué par M. le Dr. Sorel, Médecin général inspecteur délégué de l'Afrique Occidentale Française. Supplément au Bulletin mensuel 30 (2), février, 1938. Paris, 1938. 154 pp. RA 421.03

This is a second report on the analysis of French colonial vital statistics which has been carried on regularly since 1930. It covers the period from 1930 to 1935. The expansion of the civil authority is noted, especially in the cities, although meaningful data for the interior regions are available only in places such as the Cameroons, where mobile medical units have permitted the collection of field data.

There is a special section on the colonial possessions in the Atlantic, pp. 108–127. Between 1931 and 1935 the registered birth rate for Guadeloupe fluctuated between 19.57 and 22.38, the death rate between 11.43 and 15.64. Vital rates are given for the larger divisions and communes. The total birth rates for Martinique were in the low twenties, although the birth rates in the communes of the north and the east coast were 35 or above. Other communes had registered birth rates as low as 12, 13, or 14. The general death rate was about 17, although the communal rates ranged from 4.14 to 25.49. In French Guiana, the birth rate was 20.44, although it ranged from a low of 4.78 in one commune to 39.49 in Saint-Laurent. The general death rate was 24.13, although it ranged from a low of 4.27 in the commune of Kaw to 129.29 in Saint-Laurent. No vital statistics are reported from the Territory of Inini. Numbers of births and deaths are reported from St. Pierre and Miquelon, but no rates were computed.

Part V

AMERICAN COLONIES OF THE NETHERLANDS

CURAÇAO AND SURINAM

Historical

Statistical data on the Dutch possessions in the Americas have been published for nearly a century in the annual report of the *Departement van koloniën*.[1] From approximately 1850 to 1930 the *Koloniaal verslag* was issued as a one-volume report on the Netherlands colonial empire. Since 1931 it has appeared as three two-part publications, the *Indisch verslag,* dealing with the East Indies, the *Curaçaosch verslag* and the *Surinaamsch verslag.*

Statistics on the population of Surinam have been recorded since the Dutch established a civil government in that country in 1828. Their reliability has varied considerably from one period to another, dependent not only on the administrative organization of the statistical system itself but also on the changing social and ethnic composition of the population actually resident in the colony. The interior jungles are inhabited by Indians and Bush Negroes, the descendants of escaped slaves. These groups have never been counted, although estimates of total numbers have been made annually.

Information on the balance of the population is derived from registers. Prior to 1921 these were maintained by the civil authorities in each populated place, and consisted of annual estimates of the population, each estimate based upon the previous one and computed by means of vital statistics as locally recorded. These local estimates were consolidated and published by the Colonial Office, appearing in its annual report from the first half of the nineteenth century down to 1920. Until 1863 these statistics covered only the Europeans and other groups living in concentrated settlements along the coast. In 1863 slavery was abolished, and indentured Indians and West Indians were imported for plantationlab or. The more heterogeneous and fluid character of the population made the problem of securing comprehensive statistics increasingly difficult.

In 1921 there was a special population count which appears to have had some of the characteristics of a general census.[2] No attempt was made to enumerate the interior Indians or the Bush Negroes. Since 1921, statistics on the population of Surinam have been based on this count. A population register has been maintained for the country as a whole since 1921.

Population data for Curaçao and its dependencies are similar to those for Surinam. Prior to 1930, all statistics were based on consolidations of the returns from the local authorities. The first census of the six islands of the Curaçao region—Aruba, Bonaire, Curaçao, Saba, St. Eustatius, and Dutch St. Martin—

[1] Demographic statistics were also included in various other publications of the Dutch Centraal bureau voor de statistiek. See also: Netherlands. Departement van Buitenlandsche zaken. *Handbook of the Netherlands and overseas territories.* The Hague, Govt. Printing Office, 1931. vii, 405 pp. DJ 21.A5 1931.

Benjamins, H. D., and Snellemen, Joh F., Directors Encyclopaedie van Nederlandsch West-Indië, 's-Gravenhage, Martinus Nijhoff, 1914–1917. x, 782 pp. F 2141.E53.

Commissie Nederland-Curaçao, 1634–1934. *Gedenkboek Nederland-Curaçao, 1634-1934;* Uitgegeven ter gelegenheid der herdenking van de driehonderdjarige vereeniging van Curaçao met Nederland. Amsterdam, J. H. de Bussy, 1934. 382 pp. F 2409.G35.

Amsterdam Koloniaal institut. *Suriname: sociaal-hygiënische beschouwingen.* Amsterdam, 1927. 590 pp. RC 960.A55., No. 14

[2] The results of this count were published in the *Koloniaal Verslag* for 1922 and 1923.

was taken December 31, 1930–January 1, 1931.[3] The census population was 10 percent above the population estimate for the previous year. Although this census, together with the population register established in 1929–30, was presumably to be the source of post-censal estimates, the actual estimates for 1931 and later years appear to ignore the census count. A second census was scheduled for 1940, but there is no evidence that it was taken.

POPULATION STATISTICS—HISTORICAL

Departement van koloniën.
Medeelingen betreffende de koloniën. Art 60 der grondwet. Aan den heer vorzitter der Tweede kamer van de Staten-generaal, zitting 1849–1850. 's-Gravenhage, 1850. 1 vol.
This publication contains data for 1848 similar to that published for later years in the Koloniaal verslag. JV 33.N2A4 (folio)

Departement van koloniën.
Koloniaal verslag. 's-Gravenhage, 1850–1930. Annual, JV 33.N2A4.
A different edition of the material in the Koloniaal verslag for the years 1849–1854 is bound under the title: Verslag van het beheer en den staat der Nederlandsche bezittingen en koloniën in Oost-en West Indie en ter kust van Guinea over 1849–[1854] . . . Utrecht, Kemink en zoon, 1857–58. JV 33.N2A5
The basic organization of this periodical did not change during nearly a century of publication. It consisted of three parts, the first dealing with Java and adjacent possessions, the second with Surinam, and the third with Curaçao and adjacent possessions. In the earliest volumes the statistics on population were presented with the text on population, at the beginning of each part. A number of appendixes were bound with the text of each part; these consisted of decrees or essays on some aspect of colonial administration. By 1874 the tabular information on population was divorced from the text, and was presented in various appendixes to the principal parts of the report. This publication was replaced in 1931–1932 by three two-volume publications: *Indisch verslag* (covering the Netherlands East Indies), *Curaçaosch verslag* (covering the six island areas of the Curaçao region), and *Surinaamsch verslag.*
The following population statistics are included for Curaçao: Vital statistics for Aruba, Bonaire, Curaçao, Saba, St. Eustatius, and St. Martin, including marriages, births, stillbirths, and deaths; population for each area by place of birth and religion; and in later volumes, occupational distribution, mortality by age, and births by legitimacy status.
The statistics in the Surinam section include the following: Vital statistics for the coastal population; estimates of the size of the Indian and Bush Negro population; vital statistics for the military and civilian personnel associated with the Royal Army Medical Corps; and deaths in Parimaribo by cause. Some issues presented vital statistics only for Europeans, others included data for all groups except the Bush Negroes and Indians. The results of the population count of 1921 were presented in the report for 1922 and 1923.
Immigration and emigration data, with separate information for British Indians and West Indians imported as indentured servants, are available for the entire period of publication.

Central Bureau voor de statistiek.
Maandcijfers en andere periodieke opgaven betreffende Nederland en de Koloniën. Nieuwe Volkgreeks. No. 1, jaar 1898, 's-Gravenhage, 1899. Irregular, occasionally several issues per year. HA 1383.A4
This publication occasionally included vital statistics for Surinam and Parimaribo.

Centraal bureau voor de statistiek.
Jaarcijfers voor Nederlanden . . . Annuaire Statistique pour les Pays-Bas. 's-Gravenhage, 1883–19—.
Prior to 1921, there were sections on Curaçao and Surinam which included demographic statistics. The population of each island of the Curaçao group was given by sex, country of origin, and religion, with regional summaries for ten-year periods. Some data on vital statistics and the occupational composition

---

[3] The *Curaçaosch verslag* containing the original presentation of the result of the census was not located, although data from the census were included in subsequent issues.

of the population were also included. Population data for Surinam were similar including labor force statistics by country of origin, information on industrial servants, population estimates, and vital statistics classified by age, sex and legitimacy.

POPULATION STATISTICS—CURRENT

Departement van Koloniën.

Curaçaosch verslag, 1938. I. Tekst van het verslag van bestuur en staat van Curaçao over het jaar 1937. II. Statistisch jaaroverzicht van Curaçao over het jaar 1937. 's-Gravenhage, Algemeene Landsdrukkerij, 1938. I., 84 pp. II, 383 pp. J 154.R15

Vol. I. contains the population estimates of each island as of the end of the year. Vol. II. contains the detailed reports of the Census Office and the Civil Register.

Surinaamsch verslag, 1938. I. Tekst van het verslag van bestuur en staat van Surinam over het jaar, 1937. II. Statistisch jaaroverzicht over het jaar 1937. 's-Gravenhage, Algemeene Landsdrukking, 1939. I., 86 pp. II., 198 pp. J 155.R15

Contents and general arrangement are similar to those of the *Curaçaosch verslag.*

Part VI

AMERICAN TERRITORIES AND POSSESSIONS OF THE UNITED STATES

ALASKA

Historical

The earliest recorded count of any Alaskan natives was made by an agent of a Russian fur company, in 1792, and covered the villages on Kodiak Island and one or two settlements on the mainland. Subsequent counts of parts of the Russian colonies were taken by government officials in 1812, 1818, or 1819, 1822, and 1825. The most accurate counts or estimates are considered to be those of the priest Veniaminoff. These include a statement of population fluctuations between 1781 and 1830 in the districts of Kodiak and the Aleutian Islands, a census of the Aleutians in 1831, and an estimate of the total population of Alaska in 1839, which was published in full in the Alaskan volume of the Tenth Census of the United States.[1] In 1860 the Russian Holy Synod issued a report on Christians in Russian America, based on information furnished by priests and missionaries. Several reports by government officials were issued through 1863, but these appear to be quite inaccurate. A more detailed account of Russian data may be found in the Introduction to the Alaskan volume of the Eleventh Census of the United States.[2]

The first official United States report after the acquisition of Alaska in 1867 was contained in Major General Halleck's report to the United States army. Information first appeared in the United States Census in a special report in the Tenth Census, which included population, a review of the fur trade, fisheries, minerals and agriculture, geography and topography, a historical sketch, and notes on Alaskan ethnology. The Eleventh Census also included a special volume on Alaska. The results for both 1880 and 1890 were based only partially on actual enumeration, since information that could not be secured directly was estimated on the basis of records and personal knowledge of missionary priests. Succeeding censuses have become increasingly more complete and accurate. The unusual climatic conditions, the scattered population, and the difficulties of communication make a spring enumeration practically impossible. Hence the last two censuses have been taken as of October 1 of the year preceding the official census date for the United States.[3]

Vital statistics are collected by the Registrar of Vital Statistics and are published biennially in the Report of the Auditor of Alaska. Yearly totals of births, deaths, marriages, and adoptions are included in the annual report of the governor.

---

[1] Census Office. *10th Census, 1880.* Alaska: its population, industries, and resources. pp. 25–76 in: Tenth Census. June 1, 1880. Vol. 8. Washington, Govt. Printing Office, 1884. HA 201.1880.B1

[2] Census Office. *Eleventh Census, 1890.* Census Reports. Eleventh Census, 1890. Vol. VIII. Report on population and resources of Alaska at the Eleventh Census: 1890. Washington, Govt. Printing Office, 1893. pp. ix–xi. HA 201.1890.B1

[3] Bureau of the Census. *Sixteenth Census of the United States, 1940.* Population. Vol. I. Number of Inhabitants. . . . Washington, Govt. Printing Office, 1942. p. 1191.

TERRITORIAL CENSUSES

CENSUS OF 1880

Census Office. 10th Census, 1880

Alaska: its population, industries, and resources. vi, 189 pp. "The seal islands of Alaska." 188 pp. in: *Census Reports, Tenth Census. June 1, 1880.* Vol. 8. The newspaper and periodical press. Alaska: its population, industries, and resources. The seal islands of Alaska. Ship-building industry in the United States. Washington, Govt. Printing Office, 1884. HA 201.1880.B1

Census reports. . . . Vol. I. Statistics of the population. . . . Washington, Govt. Printing Office, 1883. [Alaska, pp. 695–699.] HA 201.1880.B1

CENSUS OF 1890

Census Office. Eleventh Census, 1890.

Census reports. *Eleventh Census: 1890.* Vol. VII. Report on population and resources of Alaska at the Eleventh Census: 1890. Washington, Govt. Printing Office, 1893. xi, 282 pp. HA 201.1890.B1

Census reports . . . *Vol. I. Population.* Washington, Govt. Printing Office, 1895. [Alaska, pp. 966–968.] HA 201.1890B1.

CENSUS OF 1900

Bureau of the Census.

Census reports. *Twelfth Census of the United States, taken in the year 1900.* Washington, Govt. Printing Office, 1901–1902. HA 201.1900.B1

Vol. I. Population. Part I. 1006 pp. [Alaska is included among the states.]

Vol. II. Population. Part II. 754 pp. [Alaska is included among the states.]

Occupations at the Twelfth Census. Washington, Govt. Printing Office, 1904. cclxvi, 763 pp. HA 201.1900.B2

Abstract of the Twelfth Census of the United States, 1900. Washington, Govt. Printing Office, 1902. xiii, 395 pp. HA 201.1900 D

CENSUS OF 1910

Bureau of the Census.

Thirteenth Census of the United States, taken in the year 1910. Reports. . . . Washington, Govt. Printing Office, 1912–1913. HA 201.1910.A15

Vol. III. Population. Reports by states . . . Nebraska-Wyoming, Alaska, Hawaii, and Porto Rico. 1913. 1225 pp. [Alaska, pp. 1127–1153]

Vol. VII. Agriculture. Reports by states . . . Nebraska-Wyoming, Alaska, Hawaii, and Porto Rico. 1913. 1013 pp. [Alaska, pp. 970–973]

Vol. IX. Manufactures. Reports by states . . . 1912. 1404 pp. [Alaska, pp. 1375–1378]

Abstract of the census. Statistics of population, agriculture, manufactures, and mining . . . with supplement for Alaska. Washington, Govt. Printing Office, 1913. 659 pp. [Alaska, pp. 656–606.] HA 201.1910.A2A2

The following separate bulletins duplicate the materials cited above: Alaska, Number of inhabitants. Alaska, Composition and characteristics of the population.

CENSUS OF 1920

Bureau of the Census.

Fourteenth Census of the United States, taken in the year 1920. Reports . . . Washington, Govt. Printing Office, 1921–1923. HA 201.1920.A15

Vol. I. Population. Number and distribution of inhabitants . . . [Alaska, pp. 680–681]

Vol. III. Population. Composition and characteristics of the population . . . [Alaska, pp. 1157–1169]

Vol. IV. Population. Occupations. [Alaska, pp. 1261–1269.]

Vol. VI. Agriculture . . . Part 3. . . . [Alaska, pp. 367–371.]

Vol. IX. Manufactures . . . [Alaska, pp. 1665–1667.]

The following separate bulletins duplicate the materials citied above: Alaska, Population. Population of outlying possessions. [Reprint from Vol. I.] Occupation statistics for Alaska, Hawaii, and Porto Rico. [Reprint from Vol. IV.]

CENSUS OF 1930

Bureau of the Census.

Fifteenth Census of the United States: 1930. Reports. Washington, Govt. Printing Office, 1931. HA 201.1930.A3P6

Population. Vol. I. [Alaska, pp. 1217–1227]

Outlying territories and possessions, number and distribution of inhabitants, composition and characteristics of the population, occupation, unemployment and agriculture. [Alaska, pp. 3–34.]

Abstract of the Fifteenth Census of the United States . . . . VIII, 968 pp.

The following separate bulletins duplicate the materials cited above: Alaska, Number and distribution of inhabitants. Alaska, Composition and characteristics of the population.

CENSUS OF 1940

Bureau of the Census.

Sixteenth Census of the United States: 1940. Population. Vol. I. Number of inhabitants . . . Washington, Govt. Printing Office, 1942. 1236 pp. [Alaska, pp. 1189–1197.]

Agriculture. Territories and possessions . . . 1943. vi, 306 pp.

Manufactures, 1939. Vol. III. 1942, xii, 1192 pp. [Alaska, pp. 1189–1197.]

Manufactures, 1939. Outlying areas. 1943. Vol. III, 38 pp. [Alaska, pp. 4–9.]

First Series Population Bulletins. Number of inhabitants. Alaska 1942. 7 pp. Territory of Alaska. Population. Composition and characteristics.

Population, Characteristics of the population (with limited data on housing). Washington, Govt. Printing Office, 1943. iv, 20 pp.

Presents data on sex, age, race, linguistic stock, nativity, place of birth, citizenship, marital status, occupation, and industry, and for occupied dwelling units by tenure, number of rooms, and value or monthly rent.

CURRENT VITAL STATISTICS

Alaska [Ter.]. Office of Auditor.

Report of the Auditor of the Territory of Alaska, 1939–40. Juneau, Alaska, 1941. 32 pp. HJ 11.A423

The Auditor's Office includes the Registrar of Vital Statistics for the Territory. The section on vital statistics in this biennial report includes a discussion of delayed registration of births, and statistics on births, deaths, cause of deaths, marriages, and adoptions for 1939 and 1940, by divisions.

Alaska [Ter.]. Governor.

Annual report of the Governor of Alaska to the Secretary of the Interior. Fiscal year ended June 30, 1942. Washington, Govt. Printing Office, 1942. 35 pp. J 87.A41

The section by the Auditor of Alaska contains total figures for births, deaths, marriages, and adoptions.

PANAMA CANAL ZONE

Historical

There is no accurate population information for the Panama Canal Zone prior to its acquisition by the United States in 1903. The Republic of Panama did not take its first national census until 1911, although the country was included in the Colombian census of 1870.

The first United States census of the Canal Zone was taken on February 1, 1912, by the Department of Civil Administration of the Isthmian Canal Commission. The Canal Zone was included in the regular decennial censuses of 1920, 1930, and 1940, the count being taken under the supervision of the governor. A house-to-house count of the civil population was taken by the police force after the Census of 1940. The results are presented in the Annual report of the Governor of the Panama Canal, 1941.[1]

Vital statistics are published annually in the report of the Health Department, and a brief summary is given in the annual report of the Governor. These reports include the cities of Panama and Colon, over which the United States has jurisdiction in matters relating to sanitation and public health.

TERRITORIAL CENSUSES

CENSUS OF 1912

U. S. Isthmian Canal Commission, 1905–1914.

Census of the Canal Zone, February 1, 1912. Mount Hope, C. Z., I. C. C. Press, Quartermaster's Department, 1912. 58 pp. [Includes employees of the Isthmian Canal Commission and the Panama Railroad Company working in the Canal Zone, and residing in the cities of Panama and Colon.] HA 855. A 4 1912

CENSUS OF 1920

Bureau of the Census.

Fourteenth Census of the United States, taken in the year 1920 . . . Reports. Washington, Govt. Printing Office, 1921–1923. HA 201.1920.A15

Vol. I. Population. Number and distribution of inhabitants. [Panama Canal Zone, p. 691.]

Vol. III. Population. Composition and characteristics of the population . . . [Panama Canal Zone, pp. 1239–1253.]

The following separate bulletin duplicates the materials cited above: Panama Canal Zone, Population.

CENSUS OF 1930

Bureau of the Census.

Fifteenth Census of the United States. 1930. Reports. Washington, Govt. Printing Office, 1931–32. HA 201.1930.A3P6

Population. Vol. I . . . [Panama Canal Zone, pp. 1245–1248.]

Outlying territories and possessions . . . [Panama Canal Zone, pp. 319–338]

Abstract . . . Washington, Govt. Printing Office, 1933. viii, 968 pp. [Panama Canal Zone, p. 968.] HA201.1930.A 32Z3

The following separate bulletins duplicate the materials cited above: Panama Canal Zone, Number and distribution of inhabitants. Panama Canal Zone, Composition and characteristics of the population, occupation and unemployment.

---

[1] Panama Canal. Governor. Annual report of the Governor of the Panama Canal for the fiscal year ended June 30, 1941. Washington, Govt. Printing Office, 1941. pp. 69–70. TC 774.U57

CENSUS OF 1940

Bureau of the Census.

Sixteenth Census of the United States: 1940. Population. Vol. I. Number of inhabitants. Washington, Govt. Printing Office, 1942. vi, 1236 pp. [Panama Canal Zone, pp. 1215–1218.]

Agriculture. Territories and possessions . . . Washington, Govt. Printing Office, 1943. vi, 306 pp.

First series population bulletins: American Samoa . . . Panama Canal Zone . . . Number of inhabitants. Washington, Govt. Printing Office, 1941. 16 pp.

Panama Canal Zone. Population. Characteristics of the population. Washington, Govt. Printing Office, 1941. 28 pp.

This bulletin presents data on sex, age, race, school attendance, education, nativity, place of birth, dwellings, citizenship, and marital status.

VITAL STATISTICS

Panama Canal. Health Department

Report of the Health Department of the Panama Canal for the calendar year 1940. Mount Hope, C. Z., Panama Canal Press, 1941. vi, 122 pp. RA 192.A45

Death rates and causes of death, for Panama Canal employees. Death rates, causes of death, infant mortality, and birth rates, for the Canal Zone, Panama City, and Colon separately.

Panama Canal. Governor.

Annual report of the Governor of the Panama Canal for the fiscal year ended June 30, 1941. Washington, Govt. Printing Office, 1941. 136 pp. TC 774.U57

Summaries of vital statistics and information on immigration are included.

PUERTO RICO

Historical

The Spanish government took censuses of Puerto Rico in 1765, 1775, 1800, 1815, 1832, 1846, and 1857, but the original reports of these censuses were not located. Data on the population of towns from the censuses of 1846 and 1857 are reproduced in a *Memoria* of the *Comisión de estadística especial.*[1] The data for 1857 were evidently considered to be quite unreliable, since they were not included in the published reports of the Spanish census for that year. The censuses of 1860, 1877, and 1887 were included in the regular Spanish census volumes. The last Spanish census of the island was taken in 1897, but the data were only partially tabulated. A total population figure from this census is given in Bulletin No. 1 of the Census of 1899.[2]

The census of 1899, the first United States census of the island, included population and agriculture. Since 1910 the census of Puerto Rico has been included in the regular decennial census of the United States. A special census, taken by the Puerto Rico Reconstruction Administration in 1935, included only population and agriculture.

Vital statistics are published by the Health Department at San Juan, in monthly bulletins and in an annual report. Puerto Rico was admitted to the death-registration area in 1932. Annual figures on deaths by age, race, and cause are included in the supplement to the vital statistics reports of the United States.

TERRITORIAL CENSUSES

CENSUS OF 1860

Spain. Instituto geográfico y estadístico.

Censo de la población de España según el recuento verificado en 25 de diciembre de 1860 por la Junta general de estadística. Vol. I. Madrid, Imprenta nacional, 1863. xiv, 819, xc pp. HA 1542.1860

Puerto Rico. Clasificación de los habitantes por naturaleza, sexo, estado civil y edad, pp. 775–787. Resumen de la isla por departamentos, pp. 790–793. Profesiones, artes y oficios, p. 797.

Selected statistics on the populations of towns are reproduced from the censuses of 1846 and 1857.

CENSUS OF 1877

Spain. Instituto geográfico y estadístico.

Censo de la población de España, según el empadronamiento hecho en 31 de diciembre de 1877 por la Dirección general del Instituto geográfico y estadístico. Tomo I. Madrid, Imprenta de la Dirección general del Instituto geográfico y estadístico, 1883. xxxv, 839 pp. HA 1542.1877

Isla de Puerto Rico. Su división en ayuntamientos por partidos judiciales, p. 695. Resultados generales del Censo de cada departamento por ayuntamientos, pp. 696–701. Resumen general por departamentos, pp. 702–705.

[1] Puerto Rico. Comisión de estadística especial. *Memoria referente a la estadística de la isla de Puerto Rico,* expresiva de las operaciones practicadas para llevar a cabo el censo de población que ha tenido lugar en la noche del 25 al 26 de diciembre de 1860. Puerto Rico, Establecimiento tipográfico de D. I. Guasp, 1861. 68 pp. F 1958.P87

[2] U. S. War Department. Porto Rican Census. *Census of Porto Rico,* taken under the direction of the War Department, U. S. A. Bulletin No. 1. Total population by departments, municipal districts, cities, and wards. Washington, Govt. Printing Office, 1900. p. 5 HA 902.1899.B

CENSUS OF 1887

Spain. Instituto geográfico y estadístico.

Censo de la población de España, según el empadronamiento hecho en 31 de diciembre de 1887, por la Dirección general del Instituto geográfico y estadístico. Tomo I. Madrid, Imprenta de la Dirección general del Instituto geográfico y estadístico, 1891. xiv, 920 pp. Dept. Comm.

Censo de la isla de Puerto Rico. Resultados generales. Población de hecho y de derecho, clasificada con distinción de color, por sexo, estado civil, instrucción elemental y edad. pp. 773–787.

CENSUS OF 1899

U. S. War Department. Porto Rican Census.

Report on the Census of Porto Rico, 1899. Washington, Govt. Printing Office, 1900. 417 pp. HA 902.1899B

Spanish edition has title: *Informe sobre el censo de Puerto Rico, 1899.* Washington, Imprenta del gobierno, 1900. 413 pp. Dept. Comm.

CENSUS OF 1910

Bureau of the Census.

Thirteenth Census of the United States, taken in the year 1910. Reports . . . Washington, Govt. Printing Office, 1912–1913 HA 201.1910.A15

Vol. III. Population. Reports by states . . . Nebraska-Wyoming, Alaska, Hawaii and Porto Rico. 1913. 1225 pp. [Porto Rico, pp. 1179–1225]

Vol. VII. Agriculture. Reports by states . . . Nebraska-Wyoming, Alaska, Hawaii, and Porto Rico. 1913. 1013 pp. [Porto Rico, pp. 986–1013]

Vol. IX. Manufactures. Reports by states . . . 1912. 1404 pp. [Porto Rico, pp. 1393–1401]

Abstract of the census. Statistics of population, agriculture, manufactures, and mining for the United States . . . with supplement for Porto Rico . . . Washington, Govt. Printing Office, 1913. 659 pp. HA 201.1910.A2P2

The following separate bulletins duplicate the materials cited above: Porto Rico, Number of inhabitants. Porto Rico, Composition and characteristics of the population.

CENSUS OF 1920

Bureau of the Census.

Fourteenth Census of the United States, taken in the year 1920 . . . Reports . . . Washington, Govt. Printing Office, 1921–1923. HA 201.1920.A15

Vol. I. Population. Number and distribution of inhabitants. 1921. 695 pp. [Porto Rico, pp. 682–690.]

Vol. III. Population. Composition and characteristics of the population . . . 1922. 253 pp. [Porto Rico, pp. 1195–1219]

Vol. IV. Population. Occupations. 1923. 1309 pp. [Porto Rico, pp. 1286–1309]

Vol. VI. Agriculture . . . Part 3 . . . 1922. 423 pp. [Porto Rico, pp. 383–418]

Vol. IX. Manufactures . . . 1923. 1698 pp. [Porto Rico, pp. 1683–1694]

The following separate bulletins duplicate the materials cited above: Porto Rico, Composition and characteristics of the population. Population of outlying possessions. (Reprint from Vol. I). Occupation statistics for Alaska, Hawaii, and Porto Rico. (Reprint from Vol. IV).

CENSUS OF 1930

Bureau of the Census.

Fifteenth Census of the United States: 1930. Reports. Washington, Govt. Printing Office, 1931-1932. HA 201.1930.A3P6

Population, Vol. I. Number and distribution of inhabitants . . . 1931. 1268 pp. [Puerto Rico, pp. 1249–1263]

Outlying territories and possessions, number and distribution of inhabitants, composition and characteristics of the population, occupation, unemployment and agriculture. 1932. 338 pp. [Puerto Rico, pp. 117–253]

Abstract . . . Washington, Govt. Printing Office, 1933. viii, 968 pp. [Puerto Rico, pp. 963–964] HA 201.1930.A32Z3

The following separate bulletins duplicate materials cited above: Puerto Rico, Number and distribution of inhabitants. Puerto Rico, Composition and characteristics of the population, occupations and unemployment.

CENSUS OF 1935

Puerto Rico Reconstruction Administration.

Census of Puerto Rico: 1935. Population and agriculture. Washington, Govt. Printing Office, 1938. vi, 154 pp. HA 901.A5.1935

Chapter 1. Number and distribution of inhabitants.

Chapter 2. Characteristics of the population. Occupations and employment status.

Chapter 3. Farms and farm property, crops, and livestock.

Also issued as separate bulletins.

CENSUS OF 1940

Bureau of the Census.

Sixteenth Census of the United States, 1940. Population, Vol. I. Number of inhabitants . . . Washington, Govt. Printing Office, 1942. vi, 1236 pp. [Puerto Rico, pp. 1219–1232]

Agriculture. Territories and possessions . . . 1943. vi, 306 pp.

Manufactures. Vol. III. Reports for states and outlying territories. 1942. xii, 1192 pp. [Puerto Rico, pp. 1139–1152]

Manufactures, 1939. Outlying areas. 1943. iii, 38 pp. [Puerto Rico, pp. 25–38.]

Puerto Rico. Population. Characteristics of the population. Washington, Govt. Printing Office, 1943. 82 pp.

This bulletin presents data on the basic characteristics of the population, including sex, age, color, nativity, place of birth, citizenship, marital status, school attendance, literacy, ability to speak English, employment status, class of worker, occupation, and industry.

Puerto Rico. Occupations and other characteristics by age. Washington, Govt. Printing Office, 1943.

Data are presented in this bulletin on marital status, literacy, ability to speak English, employment status, and occupation, each cross-classified by age. Statistics on relationship to head of household, hours worked during the census week, and months worked in 1939 are also included.

Puerto Rico. Housing. General characteristics of housing. 1943. 115 pp.

This bulletin presents data on dwelling units classified by occupancy, tenure, color of occupants, value or monthly rent, type of structure, exterior material, state of repair, number of rooms, size of household, and facilities.

Puerto Rico. Agriculture. Farms, farm property, livestock, and crops. Washington, Govt. Printing Office, 1942. 84 pp.

Puerto Rico. Census of business, 1939. 6 pp.

This report is a summary of the findings of the first Census of Business of Puerto Rico. It presents statistics on retail and wholesale trade, service establishments, hotels, and places of amusement.

CURRENT VITAL STATISTICS

Government of Puerto Rico. Department of Health.

Report of the Commissioner of Health to the Hon. Governor of Puerto Rico, for the fiscal year 1940–1941. San Juan, P. R., Bureau of Supplies, Printing, and Transportation. 1942. x, 357 pp. RA 194.P8327

This annual report includes a general statistical review and sections on population, natality, stillbirths, general mortality, leading causes of death, specific causes of death, maternal and infant mortality, nuptiality, and divorces.

Government of Puerto Rico. Department of Health.

Health Bulletin, Vol. VII. No. 1. Jan., 1943. San Juan, P. R. Department of Health. 22 pp. RA 194.P83365

Vital statistics of the previous four months are published in these monthly bulletins. Statistics for September are given in the January issue. Information is given on births, marriages, deaths, and causes of death, by municipalities.

U. S. Bureau of the Census.

Vital statistics of the United States, 1940. Part I. Natality and mortality data for the United States tabulated by place of occurrence, with supplemental tables for Hawaii, Puerto Rico, and the Virgin Islands. Washington, Govt. Printing Office, 1941. iv, 657 pp. HA 203.A22 1940

VIRGIN ISLANDS OF THE UNITED STATES

Historical

Censuses of the Virgin Islands were taken at approximately 5-year intervals from 1835 to 1860, and then at approximately 10-year intervals until 1911. Those located, from 1855 on, were published in the Statistiske Meddelelser of Denmark. The census of 1850 was published as part of the Statistisk Tabelvaerk; it is probable that the three earlier censuses appeared in the same series. Total population figures from the census of 1835 are available in the Folkemaengden, 1. Februar 1901 i. Kongeriget Denmark.[1] Data for 1835, 1841, and 1846 are reproduced in the census of 1855.

The United States took a special census in 1917, including population, agriculture, manufactures, and fisheries. The islands first appeared in the regular decennial census in 1930.

The three islands comprising the Virgin Islands—St. Croix, St. Thomas, and St. John—were admitted to the birth- and death-registration areas in 1924. Vital statistics are collected by the Department of Health of St. Thomas and are included annually in vital statistics publications of the United States.

TERRITORIAL CENSUSES

CENSUS OF 1850

Resultaterne af Folketaellingen paa de dansk-vestindiske Öer den 13 Mai 1850. In: Statistisk Tabelvaerk, ny Raekke, förste Bind. [Not located. Citation taken from the Census of 1855.]

CENSUS OF 1855

Denmark. Statistiske Departement.
Folketaellingen paa de dansk-vestindiske Öer den 9 de Oktober 1855. pp. 1–48 in: Meddelelser fra det Statistiske Bureau [1 Raekke], femte Samling. Kjöbenhavn, Trykt i Bianco Lunos Bogtrykkeri, 1859. 264 pp. HA 1473.B

Denmark. Statens Statistiske Bureau.
Folketaellingen paa de dansk-vestindiske Öer den 9 Oktober 1855. 48 pp. HA 911.A2

This is a separate publication of the data cited above.

CENSUS OF 1860

Denmark. Statistiske Departement.
Folketaellingen paa de dansk-vestindiske Öer den 9 de Oktober 1860. pp. 148–204 in: Statistiske Meddelelser [2 Raekke], fjerde Bind. Kjöbenhavn, Bianco Lunos Bogtrykkeri, 1865. 243 pp. HA 1473.B

CENSUS OF 1870

The Census of 1870 was not located.

---

[1] Denmark. Statens Statistiske Departement. Folkemaengden 1. Februar 1901 i Kongeriget Denmark efter de vigtigste administrative inddelinger. Statistiske Meddelelser, fjerde Raekke, tiende Bind, tredie Haefte. København, Bianco Lunos Bogtrykkeri, 1901. 132 pp. (DC) HA 1473.B

CENSUS OF 1880

Denmark. Statens Statistiske Departement.
Folketaellingen paa de dansk-vestindiske øer den 9 de Oktober 1880. pp. 129–197 in: Statistiske Meddelelser, tredie Raekke, 6te Bind. Kjøbenhavn, Bianco Lunos Kgl. Hof-Bogtrykkeri, 1883. 239 pp. HA 1473.B

Denmark. Statens Statistiske Bureau.
Folketaellingen paa de dansk-vestindiske øer den 9 de Oktober 1880. 69 pp. HA 911.A2

This is a separate publication of the data cited above.

CENSUS OF 1890

Denmark. Statens Statistiske Bureau.
Folketaellingen paa de dansk-vestindiske Øer den 9 de Oktober 1890. pp. 323–418 in: Statistiske Meddelelser, tredie Raekke, 12te Bind. Kjøbenhavn, Bianco Lunos Kgl. Hof-Bogtrykkeri, 1892. 418 pp. HA 1473.B

CENSUS OF 1901

Denmark. Statens Statistiske Bureau.
Folketaellingen paa de dansk-vestindiske Øer den 1 Februar 1901 med et Tillaeg on Varcomsaetningen 1896/97–1901/02. Haefte V, 61 pp. in: Statistiske Meddelelser, fjerde Raekke, 12te Bind. Kjøbenhavn, Bianco Lunos Bogtrykkeri, 1903. HA 1473.B

CENSUS OF 1911

Denmark. Statistiske Departement.
Folketaellingen paa de dansk-vestindiske Øer den 1. Februar 1911. Haefte V, 45 pp. in: Statistiske Meddelelser, fjerde Raekke, en og fyrretyvende Bind. København, Bianco Lunos Bogtrykkeri, 1913. HA 1473.B

Denmark. Statistiske Departement.
Folketaellingen paa de dansk-vestindiske Øer den 1. Februar 1911. København, Bianco Lunos Bobtrykkeri, 1913. 45 pp. HA 911.A2
This is a separate publication of the data cited above.

CENSUS OF 1917

U. S. Bureau of the Census.
Census of the Virgin Islands of the United States, November 1, 1917. Washington, Govt. Printing Office, 1918. 174 pp. HA 911.A5 1917

CENSUS OF 1930

Bureau of the Census.
Fifteenth Census of the United States: 1930. Reports. Washington, Govt. Printing Office, 1931–1932. HA 201.1930.A3P6
Population. I. Number and distribution of inhabitants. . . . 1931. 1268 pp. [Virgin Islands, pp. 1265–1268.]
Outlying territories and possessions, number and distribution of inhabitants, composition and characteristics of the population, occupation, unemployment and agriculture. 1932. IV, 338 pp. [Virgin Islands, pp. 255–284.]

Abstract . . . Washington, Govt. Printing Office, 1933. viii, 968 pp. [Virgin Islands, p. 965.] HA 201.1930.A32Z3
The following bulletin duplicates materials cited above: Virgin Islands of the United States, composition and characteristics of the population, number of farms, acreage, tenure, value, mortgage debt, production, and livestock.

CENSUS OF 1940

Bureau of the Census.
Sixteenth Census of the United States: 1940. Population. Vol. I. Number of inhabitants . . . Washington, Govt. Printing Office, 1942. vi, 1236 pp. Virgin Islands, pp. 1233–1236.

Agriculture. Territories and possessions . . . Washington, Govt. Printing Office, 1943. vi, 306 pp.

Virgin Islands of the United States. *Population and Housing.* 1943. 22 pp.
Characteristics of the population for which data are presented include sex, age, race, nativity, citizenship, marital status, school attendance, illiteracy, employment status, class of worker, occupation, and industry. Housing subjects included are occupancy and tenure of dwelling units, value or monthly rent, size of household and race of occupants, type of structure, exterior material, state of repair, number of rooms, and such housing facilities and equipment as water supply, toilet facilities, bathing equipment, lighting, refrigeration, and radio.

CURRENT VITAL STATISTICS

Bureau of the Census.

Vital statistics of the United States, 1940. Part I.. Natality and mortality data for the United States tabulated by place of occurrence, with supplemental tables for Hawaii, Puerto Rico, and the Virgin Islands. Washington, Govt. Printing Office, 1941. iv, 657 pp. HA 203.A22.1940
[Virgin Islands, pp. 650–657.]

O